AF736414

Initial Results from the Fast Imaging Solar Spectrograph (FISS)

Jongchul Chae
Editor

Initial Results from the Fast Imaging Solar Spectrograph (FISS)

Previously published in *Solar Physics* Volume 288, Issue 1, 2013 and Volume 289, Issue 11, 2014

Springer

Editor
Jongchul Chae
Seoul National University
Seoul, Republic of South Korea

ISBN 978-3-319-12122-2 ISBN 978-3-319-12123-9 (eBook)
DOI 10.1007/978-3-319-12123-9
Springer Cham Heidelberg New York Dordrecht London

Library of Congress Control Number: 2014951731

© Springer International Publishing Switzerland 2015
This work is subject to copyright. All rights are reserved by the Publisher, whether the whole or part of the material is concerned, specifically the rights of translation, reprinting, reuse of illustrations, recitation, broadcasting, reproduction on microfilms or in any other physical way, and transmission or information storage and retrieval, electronic adaptation, computer software, or by similar or dissimilar methodology now known or hereafter developed. Exempted from this legal reservation are brief excerpts in connection with reviews or scholarly analysis or material supplied specifically for the purpose of being entered and executed on a computer system, for exclusive use by the purchaser of the work. Duplication of this publication or parts thereof is permitted only under the provisions of the Copyright Law of the Publisher's location, in its current version, and permission for use must always be obtained from Springer. Permissions for use may be obtained through RightsLink at the Copyright Clearance Center. Violations are liable to prosecution under the respective Copyright Law.
The use of general descriptive names, registered names, trademarks, service marks, etc. in this publication does not imply, even in the absence of a specific statement, that such names are exempt from the relevant protective laws and regulations and therefore free for general use.
While the advice and information in this book are believed to be true and accurate at the date of publication, neither the authors nor the editors nor the publisher can accept any legal responsibility for any errors or omissions that may be made. The publisher makes no warranty, express or implied, with respect to the material contained herein.

Cover illustration: Raster scan images constructed from the FISS observation of an active region on 2 June 2014. The selected wavelengths are the H?–0.40 nm, −0.07 nm, −0.03 nm, 0.0 nm, +0.03 nm, +0.07 nm (reddish color, from top to bottom, from left to right) and CaII 854.2–0.40 nm, −0.04 nm, −0.02 nm, 0.0 nm, +0.02 nm, and +0.04 nm (greenish color). The field of view is 20 arc seconds by 40 arc seconds.

Printed on acid-free paper

Springer is part of Springer Science+Business Media (www.springer.com)

Preface

The Fast Imaging Solar Spectrograph (FISS) is one of the post-focus instruments of the 1.6 meter New Solar Telescope at the Big Bear Solar Observatory. The FISS primarily aims at investigating structures and dynamics of chromospheric features. This instrument is a dual-band Echelle spectrograph optimized for the simultaneous recording of the H I 656.3 nm band and the Ca II 854.2 nm band. The imaging is done with the fast raster scan realized by the linear motion of a two-mirror scanner, and its quality is determined by the performance of the adaptive optics of the telescope.

This volume contains the paper describing the instrument and a set of papers presenting initial results obtained from the FISS. These papers illustrate the capability of the early FISS observations in the study of chromospheric features. Since the imaging quality has been much improved with the advance of the adaptive optics, one can do science better with the current FISS observations.

The FISS was made by the collaboration of the Solar Astronomy Group at Seoul National University and the Solar and Space Weather Group at Korea Astronomy Space Science Institute. Jongchul Chae, the PI, contributed to the instrument project by the basic idea, early design, and data processing, Hyungmin Park, by the software and tests, Kwangsu Ahn, by the scanner, Heesu Yang, by the grating control and software, Young-Deuk Park, as the system management, Jakyoung Nah, by the optical engineering, Biho Jang, by the mechanical and electrical engineering, and Kyung-Suk Cho, by the administration and science promotion. We greatly appreciate excellent advices from many experts including Philip R. Goode, Wenda Cao, Roy Coulter, Haimin Wang, Carsten Denker, Christoph Keller, Kiyoshi Ichimoto, Takashi Sakurai, Yoshinori Suematsu, and Saku Tsuneta.

Seoul, Republic of South Korea — Jongchul Chae
2014 July

Contents

DOI 10.1007/978-3-319-12123-9_1
Reprinted from *Solar Physics* Journal, DOI 10.1007/s11207-012-0147-x

Fast Imaging Solar Spectrograph of the 1.6 Meter New Solar Telescope at Big Bear Solar Observatory

Jongchul Chae · Hyung-Min Park · Kwangsu Ahn · Heesu Yang · Young-Deuk Park · Jakyoung Nah · Bi Ho Jang · Kyung-Suk Cho · Wenda Cao · Philip R. Goode

Received: 8 April 2012 / Accepted: 24 September 2012 / Published online: 23 October 2012
© Springer Science+Business Media Dordrecht 2012

Abstract For high resolution spectral observations of the Sun – particularly its chromosphere, we have developed a dual-band echelle spectrograph named Fast Imaging Solar Spectrograph (FISS), and installed it in a vertical optical table in the Coudé Lab of the 1.6 meter New Solar Telescope at Big Bear Solar Observatory. This instrument can cover any part of the visible and near-infrared spectrum, but it usually records the Hα band and the Ca II 8542 Å band simultaneously using two CCD cameras, producing data well suited for the study of the structure and dynamics of the chromosphere and filaments/prominences. The instrument does imaging of high quality using a fast scan of the slit across the field of view with the aid of adaptive optics. We describe its design, specifics, and performance as well as data processing

Keywords Chromosphere · Instrumentation and data management · Spectrum, visible

1. Introduction

The chromosphere is a layer of the solar atmosphere that is still poorly understood despite its importance in connecting the high-β photosphere and the low-β corona (Judge, 2006). Even though often assumed to be a hydrostatic, plane-parallel layer, the chromosphere is obviously neither hydrostatic, nor plane-parallel, and comprises a variety of inhomogeneous features that are magnetically supported against gravity and/or often move fast

Initial Results from FISS
Guest Editor: Jongchul Chae

J. Chae (✉) · H.-M. Park · H. Yang
Department of Physics and Astronomy, Seoul National University, Seoul, 151-747, Korea
e-mail: chae@astro.snu.ac.kr

Y.-D. Park · J. Nah · B.H. Jang · K.-S. Cho
Korea Astronomy and Space Science Institute, Daejeon, Korea

K. Ahn · W. Cao · P.R. Goode
Big Bear Solar Observatory, New Jersey Institute of Technology, 40386 North Shore Lane, Big Bear City, CA 92314, USA

(Rutten, 2006). It appears that a proper understanding of the chromosphere critically depends on the investigation of these features. Since these chromospheric features are usually fine-structured (down to 0.2″ or below) and short-lived (a few 10 s or shorter), and often move at high speeds (Reardon, Uitenbroek, and Cauzzi, 2009), the study of their structures, dynamics, and energetics requires imaging spectroscopy of high spectral resolution, high spatial resolution, and high temporal resolution.

One way to do imaging spectroscopy is to take a series of monochromatic images, at a number of wavelengths using a tunable filter, within a short time. In the past a tunable birefringent filter was usually used (*e.g.* Chae, Park, and Park, 2006), but now a Fabry–Pérot interferometer is preferred because of its fast tuning capability and high transmission. The imaging spectroscopy based on a Fabry–Pérot interferometer is particularly attractive in that it can do excellent imaging with a spatial resolution as high as the diffraction-limited one with the help of image processing techniques such as speckle reconstruction. The Interferometric BIdimensional Spectrometer (IBIS) of the Dunn Solar Telescope is a well-operating instrument of this kind (Cavallini, 2006). The shortcomings of this technique are the narrow free spectral range and a finite duration needed to scan a spectral line, which may make it difficult to observe high-speed features and fast-changing features, respectively.

One can do imaging spectroscopy as well by taking a series of spectrograms using a grating-spectrograph with successively changing the position of the slit across the field of view. Many instruments of this kind have been constructed, including the high-speed video spectroheliograph of the 60 cm Domeless Solar Telescope at Hida Observatory (Hanaoka, 2003). Currently this technique is not much favored since its imaging capability is poorer than the filter-based imaging spectroscopy. In contrast with the filter-based imaging spectroscopy, there is no good way to process data taken by a spectrograph to overcome the seeing problem, a big obstacle to the performance of imaging instruments on ground. For high quality imaging, the light fed into the spectrograph should be free from such troublesome seeing. Either the instrument has to be put in space or the incident light should be corrected for the seeing by, for instance, adaptive optics. Since the technology of adaptive optics is developing rapidly in solar observations, it appears that the spectrograph-based imaging spectroscopy will have satisfactory imaging capability in the near future.

The greatest advantage of a spectrograph-based imaging spectroscopy is obviously that the acquired data are primarily spectral (Hanaoka, 2003) so that it is not required to reconstruct data to obtain spectral profiles. Thus this kind of imaging spectroscopy is quite suited for the spectral study of high-speed or fast-changing features. Moreover, its wide free spectral range allows each spectrum to contain not only very strong lines, suited for the study of the chromosphere, but also weak lines and continuum, useful for the study of the photosphere, so that it becomes a straightforward task to co-align a chromospheric feature and a photospheric feature at the same location on the Sun.

A recent development in imaging spectroscopy is the integral field spectroscopy (IFS) that simultaneously takes the spectra at a number of points that correspond to image pixels in the field of view, usually with the help of fibers. This is a true three-dimensional spectroscopy in that both imaging information and spectral information are acquired simultaneously. The field of view of a typical IFS is small, but may be more suitable for study of the small-scale dynamic chromosphere revealed by *Hinode* (*e.g.* Shibata *et al.*, 2007; de Pontieu *et al.*, 2007) and other advanced ground-based telescopes (*e.g.* Langangen *et al.*, 2008), and also met with a small iso-planatic patch corrected by adaptive optics on the ground. In covering a large field of view, however, an image-scanning slit spectrograph described above may excel.

We have developed a dual-band echelle spectrograph with imaging capability, named the Fast Imaging Solar Spectrograph (FISS). This instrument was installed on a vertical table in the Coudé Lab of the 1.6 meter New Solar Telescope at Big Bear Solar Observatory (Goode *et al.*, 2003, 2010; Cao *et al.*, 2010). In this paper, we describe its concepts, details, data processing, and the performance of the early runs. More technical details and test results obtained in the development stage were previously reported by Nah *et al.* (2011) and Park (2011).

2. Basic Concepts of the FISS Design

2.1. Scientific Objectives

The major objective of the FISS is to investigate the structure and dynamics of plasma in the chromospheric features. Particularly we expect to obtain observational data which can help us in finding answers to important scientific questions such as: what is the basic mechanism of non-hydrostatic support of plasma in the chromosphere? What are the building blocks of the chromosphere? Stratified layers? Thin threads? Jet-like features? Clouds? What are the thermodynamic and dynamics properties of these features? What criteria can be used to classify a variety of jet-like features into distinct types? Which types are driven by the process of magnetic reconnection and which types are not? Is magnetic reconnection in the photosphere and chromosphere important in the heating of the chromosphere? How important are waves and shocks? How is the chromosphere dynamically and thermally linked to the photosphere and to the corona?

2.2. Concepts of Design

Our strategy to achieve these scientific objectives is to acquire observational data that can be used to determine the physical parameters – such as line-of-sight velocity, temperature, non-thermal motion, and, hopefully, electron density – of each chromospheric feature with high spatial resolution and high temporal resolution. With this strategy in mind, we have designed FISS with the following concepts. First, the FISS uses an echelle grating as the main spectral element with the aid of interference filters for the order sorting. Our adoption of this element naturally makes spectrograms – spectral-spatial maps of intensity – be the basic data to be obtained by the instrument.

Second, the instrument is able to record a properly chosen pair of two spectral bands simultaneously. This adds extra spectral information that sometimes turns out to be invaluable in the determination of physical parameters. For example, the combination of Doppler absorption width of hydrogen lines and that of Ca II lines can be used to determine temperature and non-thermal motion separately because of the difference in the atomic weight between hydrogen and calcium. This unique diagnostic is effective especially in a plasma structure – such as fibrils and prominences – that lies high above the background chromosphere and contain neutral hydrogen atoms and Ca II ions numerous enough to significantly affect both line profiles. In this kind of cloud-like structures, it is possible to distinguish the effect of the structure – such as Doppler broadening – on the observed line profiles from that of the background chromosphere with the help of a simple model of radiative transfer such as the cloud model (Beckers, 1964; Tziotziou, 2007).

Third, the instrument does imaging based on the fast scan of the slit across the field of view. This function is indispensable to the spatial identification of features. Since features may change with time, the scan should be done as fast as possible.

2.3. Data Requirement

Most of all, the instrument should be able to record any spectral band in the visible and near-infrared range. Moreover, it should allow a properly chosen pair of two spectral bands to be recorded simultaneously. The coverage of each spectral band should be wide enough (> 10 Å) to include the whole portion of a chromospheric line, the nearby continua and a few photospheric lines, and the spectral resolving power should be high enough ($> 10^5$) to make the instrumental broadening negligible even in the narrowest solar lines. We set the goal for spatial resolution to $0.2''$ at the Hα line, which is just enough to reveal fine threads, but is not so high as $0.1''$, the diffraction-limited resolution of the 1.6 meter telescope. The field of view should be large enough to contain a significant number of recognizable features, say, $60'' \times 60''$. The temporal resolution is determined by the time interval between successive scans, which in turn depends on the width of the field of view covered by each scan. We require the time interval to be shorter than 30 s. Finally, we require the data to have signal-to-noise ratios higher than 10. As a measure of the signal, we uses the rms of the spatial variation of line center contrast, which is 0.1 typically in the quiet Sun. Then our requirement is that the noise at the center of a line should be smaller than 1 % of intensity. The preliminary analysis of the feasibility of the FISS was carried out by Ahn et al. (2008) with a focus on the prominence observations.

It is not easy to satisfy all the different requirements mentioned above and conflicts may naturally arise among them: especially between high temporal resolution requirement and high signal-to-noise ratio requirement, and between high temporal resolution requirement and the wide field of view requirement. Therefore it is unavoidable to make compromise in some cases. For example, we may have to increase the exposure time for high signal-to-noise ratio at the sacrifice of either temporal resolution or field of view.

3. Details of the Instrument

3.1. Pre-slit Parts

Figure 1 schematically shows the pres-slit optical layout for the FISS. These pre-slit parts determine the imaging capability of the instrument. A beam of sunlight is gathered by the

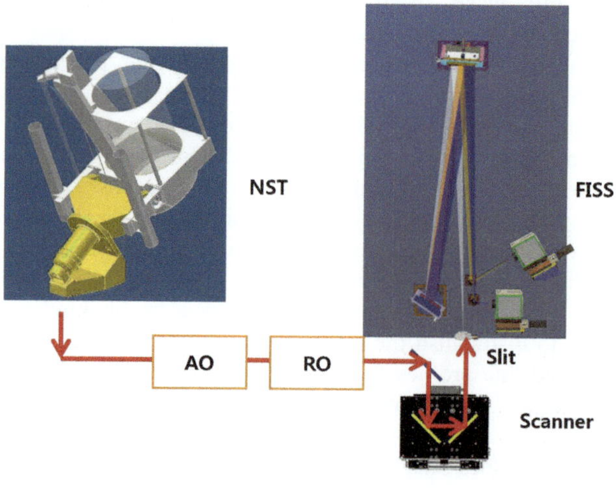

Figure 1 Schematic drawing of the optical layout of the instrument. The beam on the entrance slit is telecentric.

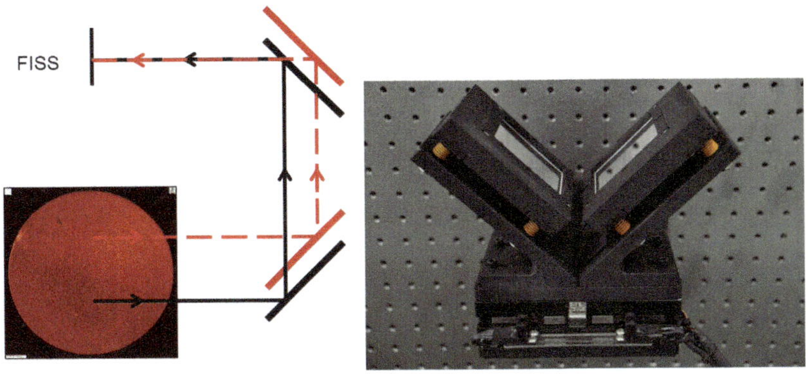

Figure 2 The optical layout (left) of the two-mirrored field scanner (right).

New Solar Telescope (NST), and is corrected for the troublesome seeing by the adaptive optics (AO). The beam is then relayed by the relay optics (RO) to the scanner and then to the entrance slit of the FISS where it becomes focused.

The NST is an off-axis Gregorian telescope with a parabolic primary mirror of $D = 1.6$ meter clear aperture. The AO currently consists of 76 wavefront sensor sub-apertures across the entrance pupil, and, when successfully operating, provides diffraction-limited images in the near IR (NIR). The next generation AO comprising 308 sub-apertures will enable diffraction-limited imaging in the visible as well. The RO consists of three folding mirrors and two lenses. It makes the beam incident on the slit not only be focused with $F = 26$, the same value as the FISS, but also be telecentric – the chief rays are set parallel to the optical axis.

3.2. Scanner

The field scanner is a device that shifts the incident beam across the slit. We do not use a commonly used scanner based on a tip-tilt mirror since it should be located at a pupil for the incident beam to have a telecentric configuration, which is a very difficult requirement in our case. Alternatively, we use a scanner of another type as depicted in Figure 2. In this type, the beam is shifted by a set of two mirrors that linearly moves in the direction across the slit. Note that a three-mirrored version of the Dove prism was previously used as a field scanner (Stolpe and Kneer, 1998), but a two-mirrored one is not. In this specific configuration, the two mirrors are set normal to each other, and the displacement of the beam becomes twice that of the scanner.

The manufactured field scanner consists of two flat mirrors mounted on an opto-mechanical grade linear motor made by the company Justek. The motor is controlled by a PMAC Mini board made by the Delta Tau company. The accurate control of this step-wise motion is realized by implementing the technique of proportional-integral-derivative (PID) gain control. This control can reduce the positional error within 1 μm, and finishes a displacement of one step (either 8 μm or 16 μm) typically within 70 ms.

The raster scan is realized by the simultaneous control of the linear motion of the scanner and the image acquisition of the cameras. The default mode is to have the scanner move step by step. It moves one step, stops, waits for each of the cameras to take one frame, and moves to the next step. It typically takes 0.1 s to proceed one step this way. The step size is set to the pixel size of the camera or the effective pixels of the resulting two-dimensional map of intensity to be of square shape. The detailed description of the field scanner is given by Ahn (2010).

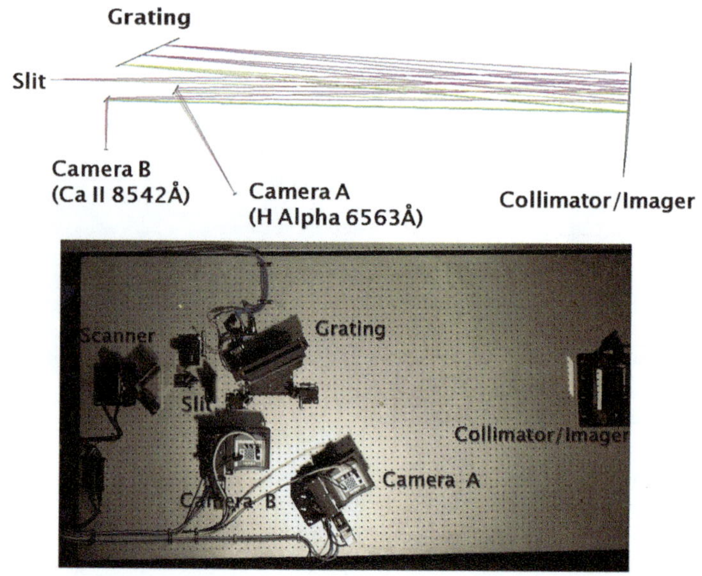

Figure 3 The optical layout (top) and the manufactured product (bottom) of the main body of the FISS. The distance between the slit and the collimator/imager is equal to the focal length of the collimator/mirror, 1500 mm.

3.3. Main Body

As is shown in Figure 3, the main body of the FISS consists of an entrance slit, a collimator/imager, an echelle grating, two broadband filters, and two cameras. The focused beam light enters the slit, becomes collimated by the collimator, gets spectrally dispersed by the grating, and then becomes focused by the imager. One spectral band is recorded on camera A, and another one, on camera B. The broadband interference filter mounted at the front of each camera transmits only the light in the spectral range of interest, removing overlapped light of other orders.

The beam is telecentric at the entrance slit, and forms a pupil at a point near the grating. To allow space for other optical components, the position of the grating was a little displaced from the pupil toward the collimator/imager. We find from the ray analysis that the effect of this deviation on the quality of imaging is negligible, and the spot diagrams are not sensitive to the precise position of the grating.

The slit has a width of $w = 32$ μm and a length of 20 mm. Currently the field of view along the slit is not determined by the slit, but by the detector chip which has a shorter length along the slit.

The mirror used as both the collimator and the imager is an off-axis paraboloid of 100 mm by 160 mm, which is a segment of a parabolic mirror of 300 mm aperture. Its focal length is $f = 1500$ mm. It was constructed in the Korea Research Institute of Standards and Science. The error in the radius of curvature is within 0.1 % and the rms error of the surface is less than $\lambda/40$. Note that the usage of the same mirror for both the collimator and the imager leads to transverse magnification of unity.

We adopted the Newport echelle grating of R2 ($\phi = 63.4^\circ$) and $\sigma = 79$ groove/mm. This grating is mounted on a rotating table, and the grating angle can be set by controlling a stepping motor using a 6K controller manufactured by Parker. The resolution of the grating

Table 1 Incidence angles for the two lines of our interest in the two optical configurations with the grating parameters: 79 grooves/mm, $\phi = 63.4°$. The parameters for the standard setup are emphasized in bold face.

Line (Å)		$\alpha - \beta = 0.93°$			$\alpha - \beta = 1.92°$		
		m	$\alpha - \phi$	B	m	$\alpha - \phi$	B
H I	**6562.8**	**34**	**−1.15**	**0.43**	34	−0.65	0.43
Ca II	**8542.1**	26	−1.65	0.42	**26**	**−1.14**	**0.43**

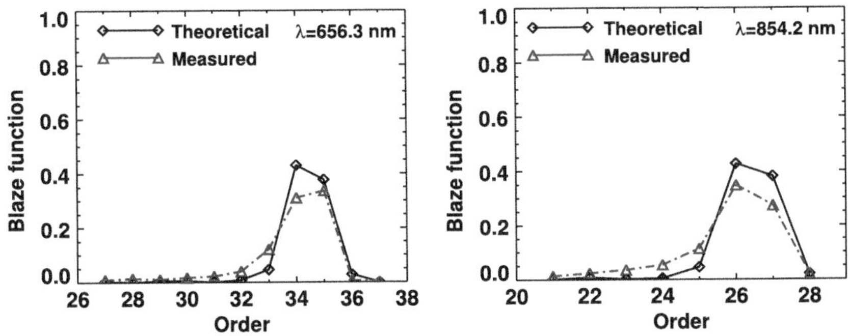

Figure 4 Theoretical and measured blaze functions for the Hα light (left) and Ca II 854.2 nm light (right), respectively. The measured values were normalized so as to make their summation be equal to the summation of all the theoretical values in each wavelength.

angle is 1.97″ per step. For the 34th order Hα light, this corresponds to the shift of 16 mÅ or one pixel on the detector.

The optical setup of each camera is determined by the value of $\theta \equiv \alpha - \beta$, the difference between incidence angle α and diffraction angle β. We choose $\theta = 0.93°$ for camera A, and $\theta = 1.92°$ for camera B. With wavelength λ given, α is to be determined from the grating equation,

$$\sin\alpha + \sin(\alpha - \theta) - \frac{m\lambda}{\sigma} = 0, \tag{1}$$

where the order m is an integer with which α maximizes the blaze function

$$B = \mathrm{sinc}^2\left(\frac{\pi\sigma\cos\phi}{\lambda}\left[\sin(\alpha - \phi) + \sin(\beta - \phi)\right]\right) \tag{2}$$

for a given ϕ. For the detailed knowledge on spectrographs including these equations, readers may refer to the textbook by Schroeder (2000). This blaze function becomes close to a maximum when $\alpha \approx \beta \approx \phi$, that is, in the quasi-Littrow configuration. With this choice, m is given by an integer that is close to $2\sigma\sin\phi/\lambda$.

Of our particular interest is the pair of the Hα line on camera A and the Ca II line at 8542 Å on camera B. The values calculated for these lines are listed in Table 1. Since this dual-band set is the most useful to our study, we regard it as the standard setup of the FISS. Regrettably, the blaze functions of this standard setup is much less than unity. This shortage arise from our choice of a commercially available grating item, Newport echelle grating: R2 that has the same values of σ and ϕ as specified above.

Figure 4 shows that the blaze functions measured as functions of order are close to theoretical values in the Hα band and in the Ca II 854.2 nm band, respectively. It is also found

Table 2 Parameters of the FISS instrumental setup. The coordinate x refers to the variation of position in the spectral direction on the detector, y, the one in the slit direction either on the detector or on the plane of sky, and z, the variation of the position across the slit on the plane of sky. Optional values are indicated inside the brackets.

Parameter	Variable	Value
Telescope aperture	D	1600 mm
Focal ratio of the beam	F	26
Slit width	w	32 [16] μm
Collimator/Image focal length	f	1500 mm
Grating groove density	$1/\sigma$	79 mm^{-1}
Grating blaze angle	ϕ	63.4°
Broadband filter (656.3 nm) bandwidth		10 nm
Broadband filter (854.2 nm) bandwidth		25 nm
Camera A deflection angle	θ	0.93°
Camera A format	$N_x \times N_y$	512 × 256 [512 × 512]
Camera A pixel size	$\delta x \times \delta y$	16 μm × 32 [16] μm
Camera A quantum efficiency at 656.3 nm	q	0.90
Camera B deflection angle	θ	1.92°
Camera B format	$N_x \times N_y$	502 × 251 [502 × 502]
Camera B pixel size	$\delta x \times \delta y$	16 μm × 32 [16] μm
Camera B quantum efficiency at 854.2 nm	q	0.35
Scanner step size	δz	32 [16] μm
Scanner stepping time		0.07 s
Scanner returning time		5 s

that the brightness of the 34th order Hα light is about 34 % of all the Hα light, and that of the 26th order Ca II 8542 Å light is about 38 %.

The order overlapping of different wavelengths is removed by using interference filters mounted in front of the CCD cameras. The full-width at half-maximum (FWHM) bandwidth of the filter for Hα is 10 nm, and the one for Ca II 8542 band is 25 nm. The peak transmissions of both filters are about 65 %.

Camera A used to record the Hα light is a back-illuminated (thin device)14-bit CCD camera DV 887 manufactured by the Andor company. Its quantum efficiency at the wavelength of the Hα light is about 90 %. This camera originally has 512 × 512 square pixels of 16 μm size. We usually bin the pixels by a factor of 1 × 2 so that the camera effectively has 512 × 256 pixels of 16 μm × 32 μm size. Since camera B recorded the Ca II line at 8542 Å is a front-illuminated (thick device) 14-bit camera DV885 made by the same company. The quantum efficiency of this camera at 8542 Å is about 35 %, being much lower than the above value. Note that we cannot use a thin device in this wavelength regime since it produces a destructive interference pattern. The original format of the camera is 1004×1002 square pixels of 8 μm size. We bin these pixels as well by a factor of 2 × 4 so that the camera effectively has 502 × 250 pixels of 16 μm × 32 μm size.

Table 2 summarizes the parameters of the FISS described so far. Note that the optional choice of the narrower slit and the finer samplings along the slit direction, and the smaller step size is to be made for higher spatial resolution observations.

3.4. System Control

A single computer controls the data acquisition by each of the two cameras, the focusing of each camera, the opening and closing of the slit cover, the rotation of the grating, and the linear motion of the scanner. The control software was developed using the Labwindows/CVI by National Instruments. In addition, the software displays data in different forms: spectrograms, raster images, line profiles, histograms of intensity, and so on. The software obviously provides the users with an interactive interface with which the users can set the operation mode, the CCD exposure times, the number of scans, scan time, step size, grating angle, and so on.

4. Data Acquisition and Processing

4.1. Data Acquisition

Three kinds of data are taken by the instrument: science data, dark/bias data, and flat/calibration data. The observing parameters for science data include the reference pixels of specified wavelengths, slit width, effective pixel size, integration times, step size, the total number of steps in each scan, and the total duration of each set of observations. Table 3 illustrates the parameters of a typical observation and the acquired data.

Dark/bias data are taken whenever the integration times are newly set. One hundred frames of dark/bias image are acquired by each camera with the same integration time as used for science data. The frames are averaged and only the average frame is finally stored

Table 3 Parameters of a typical observation and acquired data.

Observing parameters		
Slit width	w	32 μm
Camera A central wavelength	λ_c	656.28 nm
Camera A binning		1×2
Camera A exposure time	δt	30 ms
Camera B central wavelength	λ_c	854. 21 nm
Camera B binning		2×4
Camera B exposure time	δt	30 ms
Step size	δz	32 μm
Step number	N_z	100
Scan number		200
Data parameters		
Camera A data format		$512 \times 256 \times 100$
Camera A data pixel		1.9 pm $\times$ 0.16″ $\times$ 0.16″
Camera A data coverage		0.97 nm $\times$ 41″ $\times$ 16″
Camera B data format		$502 \times 250 \times 100$
Camera B data pixel		2.6 pm $\times$ 0.16″ $\times$ 0.16″
Camera B data coverage		1.31 nm $\times$ 40″ $\times$ 16″
Observing cadence	Δt	18 s
Total observing duration		1 h

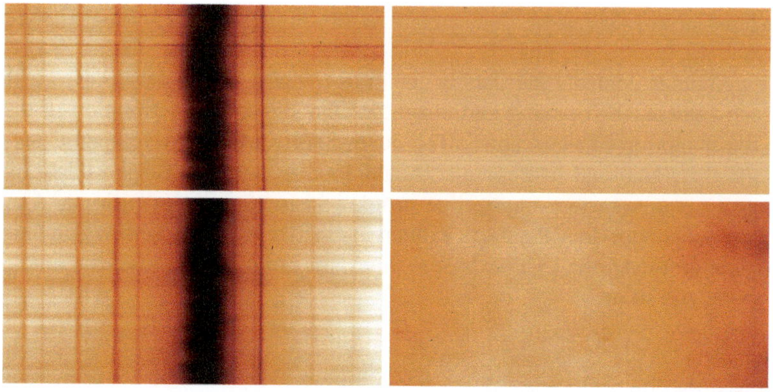

Figure 5 Left: examples of a raw Hα spectrogram (top) and its processed (bottom). Right: examples of the slit pattern (top) and the non-slit pattern (bottom) of the flat pattern. The data were taken on 23 July 2010.

into the hard disk of the computer. Subtracting this average frame does not lead to the amplification of noise that would arise when a single noisy frame of dark/bias image instead is used.

Flat/calibration data are taken more than once every observing day. A quiet region near the disk center is scanned for the purpose of flat fielding and other calibrations. The grating angle is set to ensure that spectral line of our interest is located at a specified position on the detector. The quiet region is scanned and as a result a total number of about 400 frames are acquired. These frames are averaged and only the average frame is kept for storage. Then the spectral line is displaced along the wavelength direction on the detector, and another average frame of the scan is prepared. This process is repeated seven times so that the positions of the spectral line well cover the wavelength dimension of the detector. The final products are two stored sets of seven averaged spectrograms for the two cameras. Note that one of the seven spectrograms in this flat/calibration data is used for the wavelength calibration of science data so that it should have the same spectral configuration as science data.

4.2. Flat Fielding

Correcting spectrograms for the non-uniform pattern of system response is a non-trivial part of data processing. We find it convenient to decompose the flat pattern into the slit pattern and the non-slit pattern. Figure 5 presents examples of the slit pattern and the non-slit pattern as well as a raw spectrogram.

The slit pattern originates from the non-uniformity of effective width of the slit, and is characterized by horizontal stripes. Sometimes these stripes may be tilted slightly from the horizontal direction. By taking into account this tilt, we determine this slit pattern from the median spectrogram constructed from the stored seven average spectrograms. Specifically, we obtain the two cuts of intensity along the slit direction, one at the left end of the detector, and the other at its right end. The displacement maximizing the cross-correlation of the two intensity cuts leads to the value of tilt, and the mean of the two cuts provides the one-dimensional slit pattern. By replicating this pattern in the wavelength direction with the tilt taken into account, we finally obtain the two-dimensional slit pattern. We divide the seven average spectrograms by this slit pattern and are ready to determine the non-slit pattern.

The non-slit pattern corresponds to the normal flat pattern of imaging observations. The flat pattern, in principle, is the image of a uniformly illuminating object. Since such an object does not exist on the sun, one may continuously change the field of view during the exposure so as to smooth out all the features. This simple technique, however, cannot be applicable to spectrograms, since spectral features are not smoothed out along the wavelength direction. Even though the positions of the lines on the detector are varied during the exposure, strong lines are hardly smoothed out.

Note that the proper determination of the flat pattern requires the information on its spatial variation in the two orthogonal directions: along the slit direction, that is, along the spatial direction, and across the slit direction, that is, along the spectral direction. The information on the variation of the flat pattern along the spatial direction comes from each of the seven spatially averaged spectrograms, because the spatially averaged spectrograms are supposed not to display any intrinsic spatial variation. The information along the spectral direction comes from the combination of all the spectrograms that have been relatively shifted one another along the spectral direction. To extract such information, we have developed an algorithm as described in Appendix A. This algorithm finds a pattern that optimally fits a model and is similar to the one developed by Chae (2004) for the application of filtergraph observations.

4.3. Distortion Correction

Top left panel of Figure 5 indicates that the spectral lines in the raw spectrogram are not perfectly straight, but slightly curved. This distortion results from the slight deviation of the incidence grating angle α from the value at the center of the grating, which in turn may be attributed to the slight inclination of the chief rays from the normal to the direction perpendicular to the slit. Fortunately, correcting for this distortion can be easily made only if the amount of displacement Δ_j is known. It can be determined from the cross-correlation of the j-th row spectrum and the reference spectrum that is taken from the central row. We suppress high frequency noise in Δ_j at each point by applying the piecewise quadratic polynomial fit to 30 data points surrounding the point of interest. Once Δ_j is determined, it is straightforward to correct for the distortion by shifting each row back by the amount of Δ_j. As a consequence, the spectral lines become vertically straightened.

4.4. Compression and Noise Reduction

We found the method of the principal component analysis (PCA) is very effective in reducing the amount of data without losing much physical information while suppressing noises. In the PCA, a spectral profile is expressed as a linear combination of all the principal components, and can be approximated by a linear combination of a small number of principal components. This PCA compression turns out to be successful especially when spectral profiles are similar to one another (Rees *et al.*, 2000), which is usually the case in solar observations. The details of the PCA compression used are described in Appendix B.

Let us illustrate how we have determined the set of eigenvectors and a suitable value of the compression factor N/n where N is the total number of data points in the spectral domain, and n is the number of coefficients retained in the PCA compression. Suppose a raster scan of a region is achieved by 100 steps of exposure on camera A of 512×256 pixels. It then follows that $N = 512$, and each scan contains a total of 100×256 independent spectral profiles of the Hα line. We randomly select 2000 spectral profiles to construct the covariance matrix. The eigenvalues and eigenvectors of this matrix are determined by applying an algorithm using Householder reductions and the QL method with implicit shifts. The eigenvalues and eigenvectors are placed in the descending order of eigenvalues.

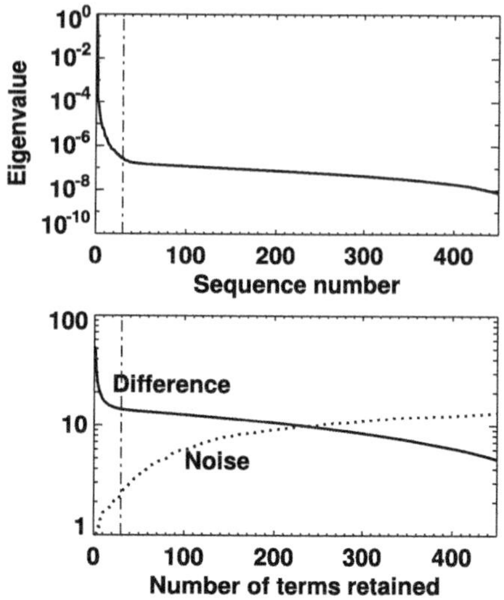

Figure 6 Eigenvalues in the decreasing order (top), and the rms differences as a function of the number of terms retained in the compression (bottom).

The top panel of Figure 6 shows the plots of eigenvalues normalized by the biggest one. It is clear from the figure that eigenvalues drop rapidly with sequence number $n < 50$ and then decrease slowly with n, with most eigenvalues being much smaller than the biggest one. We have compared a spectrogram with its compressed versions with different values of n, and have plotted the root-mean-squares of difference between the two as a function of n in the bottom panel of the figure.This difference measures the amount of information lost by the compression. It is likely that this lost information consists of more noise than statistically meaningful signal while the compressed data include more of such signal than noise. This means that the compression process naturally results in noise reduction.

We have empirically determined the standard noise in a set of data in the following way. If the sampling is fine enough to represent real features as smooth variations of intensity, pixel-to-pixel variations may be considered to mostly reflect noise. Base on this belief, we infer noise from a set of second-order finite differences each of which is calculated from three neighboring pixels. This method of estimate can be applied not only to raw data, but also to processed data.

The noise estimated in the continuum as a function of n is presented in the figure. Obviously, the smaller n is, the smaller the noise is, indicating that the compression reduces the noise. The noise in the uncompressed data is estimated to be 14 DN (data number). We choose 30 for the value of n for an optimal compression. This means that data are compressed by a factor of about 17. With this choice, the value of the rms difference is found to be about 14 DN, which is very close to the noise in the uncompressed data, and the noise is estimated to be about 1.9 DN. Therefore we conclude that the choice $n = 30$ is big enough for retaining most of non-noise features, and small enough for significantly reducing noise. A good thing is that the specific value of n is not critical to the behavior of data compression: the figure indicates that values of n in the range from 20 to 40 yields similar values of rms difference. Figure 7 illustrates the comparison of an uncompressed spectral profile and the one compressed with $n = 30$, and the profile of difference between them.

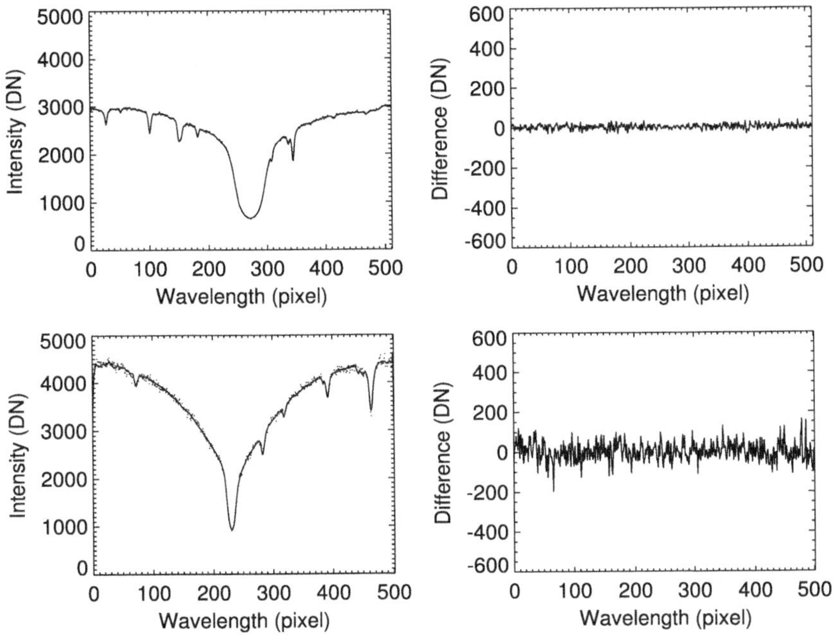

Figure 7 Left: the comparison of an compressed spectral profiles (solid) with the uncompressed ones (dots) for the Hα band and the Ca II 8542 band. Right: the profiles of difference between the two. The data were taken on 23 July 2010.

Table 4 Measured intensities and noises.

Parameter	Variable	Hα band	8542 band
Signal of continuum (measured)	S (DN)	2890	4210
Noise (uncompressed data)	N (DN)	14	46
Noise (compressed data)	N (DN)	2.8	8.3
Signal of line center (measured)	S (DN)	707	980
Signal of a 10 %-contrast feature (estimated)	S (DN)	71	98
Noise (uncompressed data)	N (DN)	7	28
Noise (compressed data)	N (DN)	1.1	4.3

5. Performance of the Instrument

5.1. Signal-to-Noise Ratio

Table 4 lists the values of light intensity and noise measured from the data. The amount of light detected by each pixel of the CCD camera is stored in unit of DN (data number). From the same set of data that is used for Figure 7, we found that the quiet Sun average of the continuum light in the Hα band is about 2890 DN, and that of the continuum in the Ca II 8542 band is about 4210 DN. We also estimated the standard noises from the uncompressed data and the PCA compressed data, respectively, using the method of second-order finite differences mentioned above. Since the detection of a chromospheric feature is

Table 5 Theoretical calculations of spectral resolution when the slit width is set to either 16 µm or 32 µm.

Parameter	Hα band	Ca II 8542 band	HeNe laser
Wavelength	6563 Å	8542 Å	6328 Å
Order	34	26	36
Detector size	19 mÅ	25 mÅ	15 mÅ
Detector coverage	9.7 Å	12.9 Å	7.8 Å
Grating resolution	19 mÅ	33 mÅ	18 mÅ
Spectral purity	18/36 m Å	24/48 mÅ	17/34 mÅ
Spectral resolution (net)	33/45 mÅ	48/64 mÅ	29/42 mÅ
Spectral resolving power	199 000/143 000	176 000/133 000	216 000/150 000

mainly through the core of these strong lines, we also determined intensities and noise at the line centers.

The values in Table 4 can be exploited to investigate the issue of signal-to-noise ratio. For instance, suppose we intend to study a feature which appears darker than the average Sun by 10 % at the central wavelength of the Hα line. The signal of this feature is then calculated to be $0.1 \times 707 = 71$ DN. Meanwhile the noise of the uncompressed data was measured to be 7 DN at the same wavelength, so that the resulting signal-to-noise ratio of the raw data turns out to be about 10. The PCA compression suppresses the noise significantly, which was found to be 1.1 DN. Therefore, the signal-to-noise ratio of the PCA compressed data has a higher value of 65. Meanwhile, in the Ca II 8542 Å line, the signal is 98 DN, and the measured noise is 28 DN, so that the signal-to-noise ratio is found to be 3.5. The PCA-compressed data are found to have a lower noise of 4.4 DN, leading to a higher signal-to-noise ratio of 22.

In Section 2.2, we mentioned that data should have signal-to-noise ratios higher than ten. Without compression, this requirement is barely achieved by spectral data of Hα line, but not by those of the Ca II 8542 Å line. Since the noise in the uncompressed data is dominated by photon noise, the low signal-to-noise ratio of the Ca II data is due to the small number of photons captured by camera B. We have investigated why the Ca II data have lower through-put than Hα data, and as a result, we found that there are two major reasons. First, the transmission of the telescope-spectrograph system is much lower in the Ca II band than in the Hα band, sine the system uses a total of 17 aluminum-coated reflecting surfaces. The ratio of the Ca II band transmission of these surfaces to the Hα band is 0.39. If all of these surfaces are silver-coated, the transmission of the system will increase by a factor of 2.9 in the Hα band, and 7.6 in the Ca II band. Another reason is obviously at the lower quantum efficiency of the front-illuminated CCD camera B, 0.35, which is much lower than that of the back-illuminated CCD camera B, 0.90.

5.2. Spectral Resolution and Spatial Resolution from a Lab Test

Table 5 presents how we calculated theoretical values of spectral resolution of the spectrograph at a few wavelengths. At the wavelength of 6328 Å, for example, the choice of the 16 µm slit leads to the spectral purity of 17 mÅ and the width of the illuminated part of the grating, to the grating resolution of 17 mÅ. The combined effect of these smearing factors as well as the finite size of a pixel, 15 mÅ, results in the theoretical net spectral resolution of 39 mÅ and the corresponding spectral power of 216 000 at the wavelength of 6328 Å.

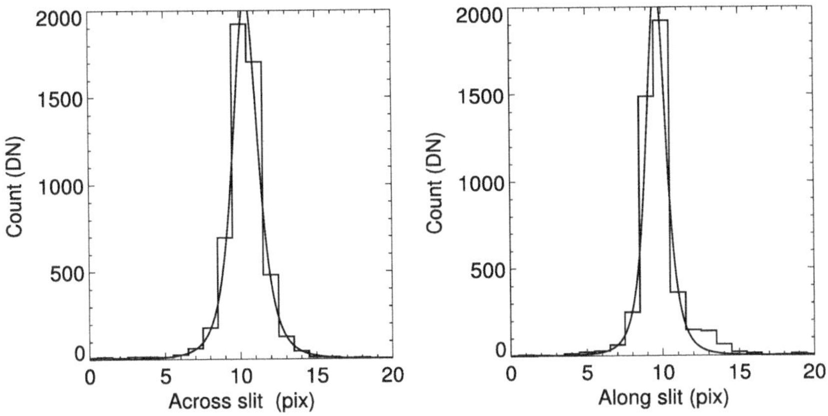

Figure 8 The spectral profile and spatial profiles of the spectral image of a monochromatic point source (laser source). The solid curves represent the model functions of the form $\sim 1/(((x-x_0)/a)^2+1)^2$.

We measured the spectral resolution and spatial resolution from a lab test using a HeNe laser source of monochromatic light at 6328 Å. The scanner was not operating this time, and the slit was set to 16 μm width. Figure 8 shows the spectral profile of the 36th order image of the source. The FWHM of the spectral profile is found to be about 2.2 pixels. With the use of the dispersion 15 mÅ/pixel at this wavelength, this value leads to the empirically determined spectral resolution of 33 mÅ and the resolving power of 1.92×10^5. These values are not much different from the values, 29 mÅ and 2.16×10^5 that we theoretically calculated.

Figure 8 also shows that the spatial profile of the point-like source has a small value of FWHM: 1.8 pixel. Since the size of a pixel is 16 μm or 0.08″, the spatial resolution of the instrument is about 0.14″, which is a measure of the performance of the imaging along the slit, and satisfactorily fulfills the requirement we set. Note that this value of spatial resolution depends on the optical setup, but neither on the performance of the scanner nor on the atmospheric seeing since it came from a lab test without using the scanner.

5.3. Imaging Capability of the Scanner

We examined the imaging capability of the scanner by taking the image of a resolution panel located at a focal position between the AO and the relay optics. The spatial resolution of this image is determined by the performance of the scanner, the optics of the combined system of the relay optics and the spectrograph, but not by the atmospheric seeing. Note that the vibration of the scanner may result in the degradation of image quality as the atmospheric seeing does; high frequency vibration may deteriorate the spatial resolution along the slit direction and low frequency vibration may introduce irregularity in the constructed raster scan image across the slit direction.

By analyzing the gradients of the blurred edges in the image of the resolution pane, we found that during the nominal operation of the scanner the FWHM *along* the slit has a value around the average 2.9 pixels (46 μm or 0.23″), which is bigger than the value obtained above with the scanner being off. On the other hand, the FWHM *across* the slit was found to have an average value of 2.2 pixels (35 μm or 0.18″). The stability of the scanning has been examined using a time series of raster images of a small quiet region constructed from the data taken by repeated scans. The cadence of the scans was fast enough to preserve the pattern of granulation. The cross-correlation analysis of two successive images taken at the

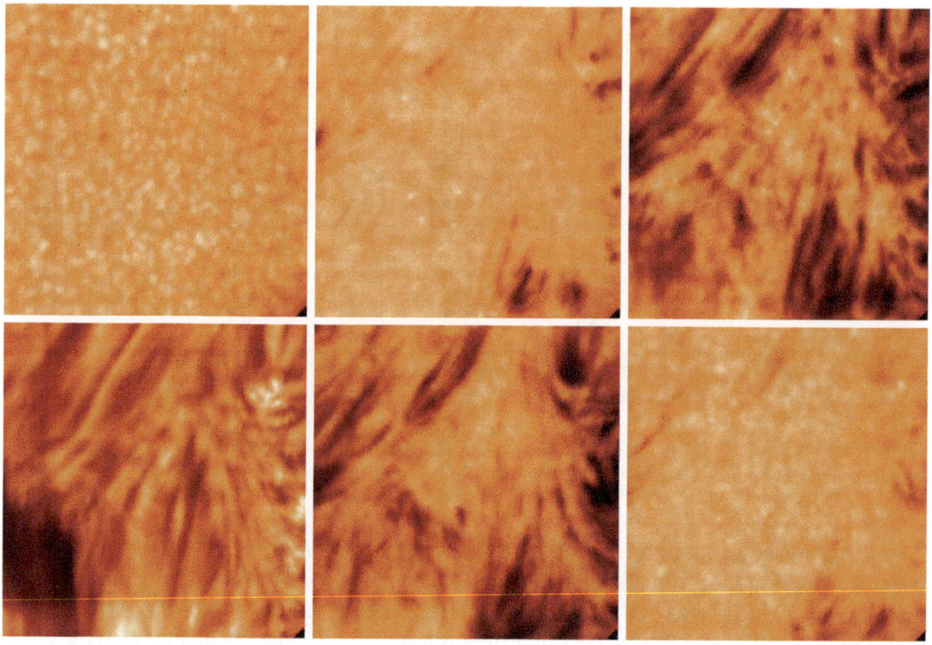

Figure 9 Raster images of a quiet region constructed at wavelengths −4.0, −0.8, −0.5, 0, 0.5, and 0.8 Å of Hα from top left to bottom right. The field of view is 38″ by 41″, and the spatial resolution is estimated to be 0.6″. The observation date was 1 July 2011.

near-continuum indicates that they cover the same field of view within the uncertainty of one step size (0.16″), indicating that the scanning of the instrument is stably repeatable.

5.4. Net Spatial Resolution of Real Observations

Generally speaking, the net spatial resolution as well as the performance of the scanning strongly depends on the atmospheric seeing condition, even when the AO operates. This is because the performance of the AO is not perfect and is sensitive to the availability of high contrast features that in turn depends on the seeing.

The images in Figures 9 and 10 represent the best ones obtained so far, one from a quiet region, and the other from an active region. They can be used to infer the net spatial resolution of the instrument that includes the effect of seeing. It is clear from the images that granulation is well identified in Hα images. This suggests that the spatial resolution of Hα observations is better than 1″. Identifying smallest bright structures, usually at the wing wavelengths, and measuring their widths along the slit direction, we obtained the estimates of spatial resolution in the these two cases, which are 0.6″ and 0.7″, respectively.

On the other hand, the Ca II 8542 observations shown in Figure 10 have poorer spatial resolution of 1.2″ than the Hα observations that were simultaneously done. The dependence of seeing on wavelength cannot explain for this discrepancy, and recently we came to realize that the focusing of early (2010 and 2011) observations had not been optimized for the Ca II band. The two lens in the relay optics are subject to chromatic abberation so that there exists a small difference in the focal position between 6563Å and 8542 Å lights. Unless the lens positions are well adjusted so that the slit becomes located between the two focal

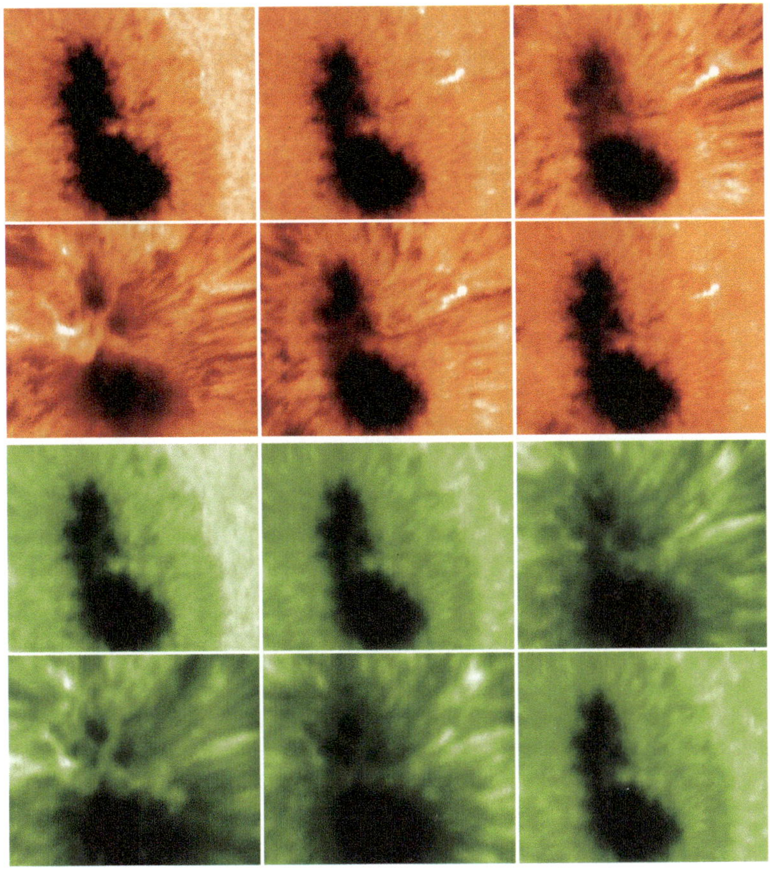

Figure 10 Top: Raster images of a quiet region constructed at wavelengths −4.0, −0.8, −0.5, 0, 0.5, and 0.8 Å of the Hα line from top left to bottom right. The field of view is 48″ by 41″, and the estimated spatial resolution is 0.7″. Bottom: Raster images of a quiet region constructed at wavelengths −4.0, −0.8, −0.3, 0, 0.3, and 0.8 Å of the Ca II 8542 line from top left to bottom right. The field of view is 48″ by 40″, and the spatial resolution is 1.2″. The data were taken on 4 August 2011.

positions, one bench becomes much out of focus even though the other bench is in focus. The chromatic abberation will be completely removed when the lenses are replaced by carefully designed achromatic ones.

5.5. Time Resolution

The number of steps in each scan determines time resolution and the width of the field of view. One step needs a total of 110 ms for the movement of the scanner (70 ms), and the exposure time (30 ms), and the spare time (10 ms). Since the readout of the cameras is done during the spare time and the scanner motion, it does not lengthen the cadence. The completion of 100 steps takes 11 s. Adding an extra time of 5 s required for the scanner to return to the origin, we finally obtain the time resolution of 16 s for successive scans each of which consisting of 100 steps. In this case, the width of the field of view is 16″. On the other hand, when the number of steps was 10, the time resolution was found to be as good as 2.1 s, and when the number of steps was 300, it was found to be as poor as 48 s.

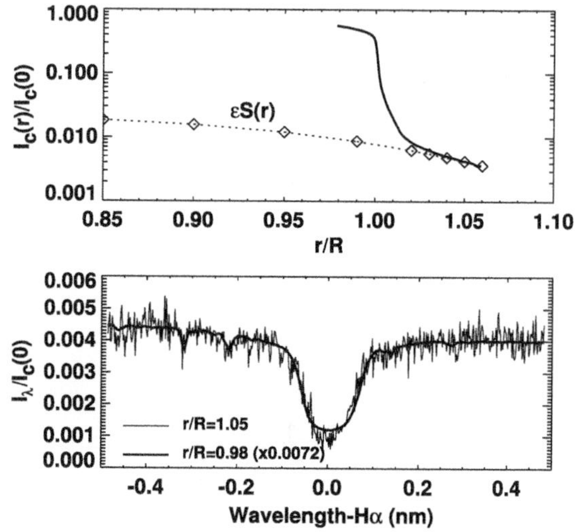

Figure 11 Top: profile of continuum intensity (solid curve) and the modeled scattered light integral (dotted curve with symbols). Bottom: Profile of the Hα line inside the limb (re-scaled by a factor of 0.0072) and the one outside the limb. The data were taken on 30 June 2010.

5.6. Stray Light

Two kinds of stray light affect the spectrograph observations: spatial stray light and spectral stray light (see Appendix C). The continuum intensity outside the solar limb is a measure of spatial stray light produced by the terrestrial atmosphere and the observing instrument. By analyzing the map of continuum intensity constructed from the spectrograms taken at the east limb on 30 June 2010, we have determined the spatial variation of the intensity across the limb as a function of angular distance from the center of the solar disk, as shown in the upper panel of Figure 11. We have modeled the scattered light integral $S(r)$ that measures the effect of large spread angle stray light by introducing a Gaussian spread function with a FWHM of $200''$ and as a result we found that the stray light fraction near disk center ϵ is about 0.027. Furthermore, the lower panel of Figure 11 supports our expectation that the effect of spatial stray light is uniform over wavelength, preserving the shape of the spectral profile. In fact, it is spectral stray light that affects it.

We estimated the fraction of spectral stray light ζ by comparing the observed spectral profile of the Hα line with the atlas of Wallace, Hinkle, and Livingston (2007). According to this atlas, the core-to-continuum intensity ratio of the Hα profile taken at the disk center of the quiet Sun is 0.16. In comparison, the core-to-continuum intensity of the average quiet Sun profile taken by the FISS is found to be 0.21. As a result, ζ is estimated to be 0.065.

6. Summary and Conclusion

We have described the design, specifics, data manipulation, and performance of the FISS. The instrument was designed to take high signal-to-noise ratio spectra of chromospheric features at high spatial resolution, high spectral resolution, and high temporal resolution, but it is not possible to meet all the requirements and it is necessary to compromise among different needs.

Figures 10 and 12 provide examples of data that are currently taken by the FISS, spectrograms that are in the original form (Figure 12) and raster images that come out of processing (Figures 10). The strength of the instrument is definitely in the capability of recording the

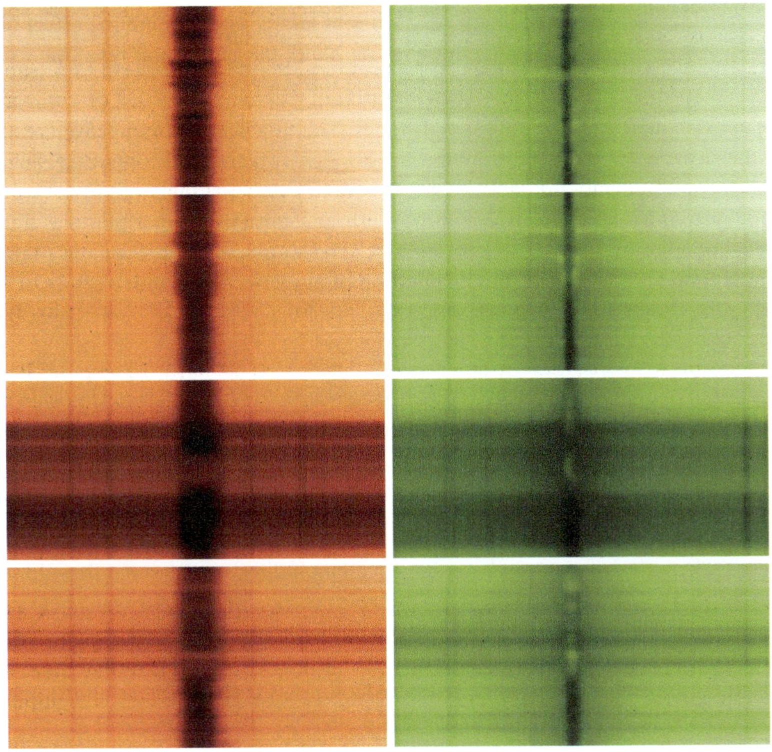

Figure 12 Examples of the pair of an Hα spectrogram and a Ca II 8542 spectrogram taken simultaneously at different locations of the active region shown in Figure 10.

two important spectral lines – Hα and Ca II 8542 Å line – exactly simultaneously. Both these lines are significantly affected by the presence of dense material in the chromosphere, but they reflect its thermal structure and dynamics in different ways. Specifically it is clear from Figure 12 that absorption structures are better reflected in the Hα line, and emission structures, better in the Ca II line.

Several important characteristics and the current performance of the instrument are summarized as follows:

i) The spectral coverage is big enough to include one strong absorption line – such as the Hα line and Ca II 8542 Å line – useful for the observation of the chromosphere and a few weaker lines useful for the observations of the photosphere.
ii) The spectral resolution is satisfactorily high. With the slit of $0.08''$ width, the spectral resolving power is above 1.5×10^5. With a wider slit, the spectral resolution is a little degraded, but is still satisfactory.
iii) We found that the PCA (principal component analysis) compression can be implemented successfully. This kind of processing not only reduces the size of data to be stored by a factor of up to 20, but also suppresses noise significantly.
iv) We found that noise in the uncompressed data is dominated by photon noise. The signal-to-noise ratio of data is hence determined by the amount of light gathered by each pixel on the CCD camera.

v) The spatial resolution of the instrument without taking into account the effect of seeing, is estimated to be about $0.25''$. On the other hand, the net spatial resolution including the effect of seeing and the AO is found to be $0.6''$ or poorer.

We plan to improve the FISS in several stages. The first stage is to improve the quality of the Ca II data by replacing the lenses in the relay optics by achromatic ones that can precisely match the focal lengths of the Hα and Ca II benches. We will also increase the throughput of the system by replacing the grating with an optimal blaze angle, and aluminum-coated reflecting surfaces by silver-coated ones. The significant improvement in the spatial resolution will be made with the implementation of higher-order adaptive optics of 308 sub-apertures in the Coudé Lab of the telescope.

Acknowledgements We are grateful to the referee for a number of constructive comments. This work was supported by the National Research Foundation of Korea (KRF-2008-220-C00022) and by the Development of Korean Space Weather Center, a project of KASI.

Appendix A: Flat Fielding

Let us denote the index of a pixel along the wavelength direction by i and that along the slit direction, by j. An averaged spectrogram, s_{ij} should depend on i only, but not on j if the spatial average is ideally done so that we may write $s_{ij} = o_i$ where o_i is the one-dimensional spectrum of the average sun. Then the kth spectrogram a^k_{ij}, that is acquired by the camera with the spectrum being displaced by x_k from the reference spectrum along the wavelength direction can be mathematically modeled as:

$$a^k_{ij} = c_k s_{i-x_k,j} f_{ij} = c_k o_{i-x_k} f_{ij}, \tag{3}$$

where f_{ij} is the flat pattern we intend to determine, and c_k is the relative level of intensity. We find it necessary to modify this equation a little bit, for the spectral lines in the acquired spectrograms are not vertically straight, but slightly curved. This distortion may be considered as a manifestation of horizontal displacement of spectrum depending on the position along the slit, which we denote by Δ_j. Taking into account this distortion, we modify the above equation into

$$a^k_{ij} = c_k o_{i-\Delta_j-x_k} f_{ij}, \tag{4}$$

or into its logarithmic form,

$$A^k_{ij} = C_k + O_{i-\Delta_j-x_k} + F_{ij}. \tag{5}$$

For given A^k_{ij}, x_k, and Δ_j, maximizing

$$H = \sum_{ijk} \left(C_k + O_{i-\Delta_j-x_k} + F_{ij} - A^k_{ij}\right)^2 w(i - \Delta_j - x_k) \tag{6}$$

and

$$K = \sum_{ijk} \left(C_k + O_i + F_{i+\Delta_j+x_k,j} - A^k_{i+\Delta_k+x_k,j}\right)^2 w(i + \Delta_j + x_k) \tag{7}$$

we obtain the iterative formulas: for F_{ij}

$$\Delta F_{ij} \approx -\frac{\sum_k (C_k + O_{i-\Delta_j-x_k} + F_{ij} - A^k_{ij}) w(i - \Delta_j - x_k)}{\sum_k w(i - \Delta_j - x_k)}, \tag{8}$$

for C_k

$$\Delta C_k = -\frac{\sum_{ij}(C_k + O_{i-\Delta_j-x_k} + F_{ij} - A^k_{ij})w(i-\Delta_j-x_k)}{\sum_{ij} w(i-\Delta_j-x_k)}, \tag{9}$$

and for O_i

$$\Delta O_i \approx -\frac{\sum_{jk}(C_k + O_i + F_{i+\Delta_j+x_k,j} - A^k_{i+\Delta_k+x_k,j})}{\sum_{jk} w(i+\Delta_j+x_k)}. \tag{10}$$

Note that $w(x)$ and $v(y)$ are functions that have values of either unity or zero depending on whether x and y are inside the corresponding index ranges, respectively. These expressions for the iteration are similar to those given by Chae (2004) so we do not go into further details.

Appendix B: Compression

Generally speaking an arbitrary vector I_i of N elements ($i = 1, \dots, N$) can be written as a linear combination of a set of N orthonormal vectors s_i^k ($k = 1, \dots, N$). Specifically we choose the form

$$I_i = \bar{I} \sum_{k=1}^{N} c_k s_i^k, \tag{11}$$

where $\bar{I}$ is the average intensity. The coefficient c_k is determined from I_i:

$$c_k = \sum_{i=1}^{N} \frac{I_i}{\bar{I}} s_i^k. \tag{12}$$

If s_i^k can be placed in descending order of statistical significance, we can reduce the number of coefficients to be kept to $n < N$ while retaining most physical information:

$$I_i \approx \bar{I} \sum_{k=1}^{n} c_k s_i^k. \tag{13}$$

Data are thus compressed by a factor of N/n.

The set of basis vectors s_i^k is constructed from the database of model profiles, J_i^l ($l = 1, \dots, M$), that well represent spectral profiles to be analyzed. By referring to Rees *et al.* (2000), we define a symmetric covariance matrix

$$A_{ij} = \sum_{l=1}^{M} \frac{J_i^l}{\bar{J}^l} \frac{J_j^l}{\bar{J}^l} \tag{14}$$

and identify its kth eigenvector with the kth basis vector s_i^k while the eigenvectors are sorted in the descending order of the associated eigenvectors. Note that the statistical significance of s_i^k is measured by its associated eigenvalue λ_k.

Appendix C: Stray Light Model

We model the effect of spatial stray (large-angle spread) light on the data using the equation

$$\frac{I_c^{\text{obs}}(\mathbf{r})}{I_c^{\text{obs}}(0)} = (1-\epsilon)\frac{I_c(\mathbf{r})}{I_c(0)} + \epsilon S(\mathbf{r}) \tag{15}$$

and that of spectral stray (large-wavelength spread) light using the equation

$$\frac{I_\lambda^{\text{obs}}(\mathbf{r})}{I_c^{\text{obs}}(\mathbf{r})} = (1-\zeta)\frac{I_\lambda(\mathbf{r})}{I_c(\mathbf{r})} + \zeta V_\lambda, \tag{16}$$

where $I_c^{\text{obs}}(\mathbf{r})$ and $I_c^{\text{obs}}(0)$ are the observed intensities of the continuum at an arbitrary position $\mathbf{r}$ and the disk center, respectively, and $I_\lambda^{\text{obs}}(\mathbf{r})$ and $I_\lambda^{\text{obs}}(0)$ refer to those at an arbitrary wavelength λ. The intrinsic intensities $I_c(\mathbf{r})$, $I_c(0)$, $I_\lambda(\mathbf{r})$, and $I_\lambda(0)$ are defined in the same way.

In our model, the effects of spatial stray light and spectral stray light on a spectrogram are independent from each other; the spatial stray light affects the level of intensity equally over different wavelengths, and the spectral stray light affects the shape of each spectrum. The spatial stray light integral $S(\mathbf{r})$ is a normalized function slowly varying over the position $\mathbf{r}$, but may be approximated to be equal to a constant value, 1, near the disk center. In a similar way, the spectral stray light integral V_λ is a normalized function slowing varying over wavelength λ, but may be approximated to be equal to a constant value, 1. Therefore, the correction of stray light in disk observations can be done if two parameters, the fraction of spatial stray light ϵ and the fraction of spectral stray light ζ, are known.

References

Ahn, K.: 2010, Development of New Astronomical Instruments for High Resolution Solar Observations. Ph.D. thesis, Seoul National University.

Ahn, K., Chae, J., Park, H.-M., Nah, J., Park, Y.-D., Jang, B.-H., Moon, Y.-J.: 2008, *J. Korean Astron. Soc.* **41**, 39.

Beckers, J.M.: 1964, A Study of the Fine Structures in the Solar Chromosphere. Ph.D. thesis, Utrecht University.

Cao, W., Gorceix, N., Coulter, R., Ahn, K., Rimmele, T.R., Goode, P.R.: 2010, *Astron. Nachr.* **331**, 636.

Cavallini, F.: 2006, *Solar Phys.* **236**, 415.

Chae, J.: 2004, *Solar Phys.* **221**, 1.

Chae, J., Park, Y.-D., Park, H.-M.: 2006, *Solar Phys.* **234**, 115.

de Pontieu, B., McIntosh, S., Hansteen, V.H., Carlsson, M., Schrijver, C.J., Tarbell, I.M., *et al.*: 2007, *Publ. Astron. Soc. Japan* **59**, 655.

Goode, P.R., Denker, C.J., Didkovsky, L.I., Kuhn, J.R., Wang, H.: 2003, *J. Korean Astron. Soc.* **36**, S125.

Goode, P.R., Coulter, R., Gorceix, N., Yurchyshyn, V., Cao, W.: 2010, *Astron. Nachr.* **331**, 620.

Hanaoka, Y.: 2003, In: Keil, S.L., Avakyan, S.V. (eds.) *Innovative Telescopes and Instrumentation for Solar Astrophysics*, *Proc. SPIE* **4853**, 584.

Judge, P.: 2006, In: Leibacher, J., Stein, R.F., Uitenbroek, H. (eds.) *Solar MHD Theory and Observations: A High Spatial Resolution Perspective*, *ASP Conf. Ser.* **354**, 259.

Langangen, Ø., De Pontieu, B., Carlsson, M., Hansteen, V.H., Cauzzi, G., Reardon, K.: 2008, *Astrophys. J.* **679**, L167.

Nah, J.-K., Chae, J.-C., Park, Y.-D., Park, H.-M., Jang, B.-H., Ahn, K.-S.: *et al.*: 2011, *Publ. Korean Astron. Soc.* **26**, 45.

Park, H.: 2011, Development of Fast Imaging Solar Spectrograph and Observation of the Solar Chromosphere. Ph.D. thesis, Chungnam National University.

Reardon, K.P., Uitenbroek, H., Cauzzi, G.: 2009, *Astron. Astrophys.* **500**, 1239.

Rees, D.E., López Aristle, A., Thatcher, J., Semel, M.: 2000, *Astron. Astrophys.* **355**, 759.

Rutten, R.: 2006, In: Leibacher, J., Stein, R.F., Uitenbroek, H. (eds.) *Solar MHD Theory and Observations: A High Spatial Resolution Perspective*, *ASP Conf. Ser.* **354**, 276.

Schroeder, D.J.: 2000, *Astronomical Optics*, 2nd edn., Academic Press, San Diego.

Shibata, K., Nakamura, T., Matsumoto, T., Otsuji, K., Okamoto, T.J., Nishizuka, N., *et al.*: 2007, *Science* **318**, 1591.

Stolpe, F., Kneer, F.: 1998, *Astron. Astrophys. Suppl.* **131**, 181.

Tziotziou, K.: 2007, In: Heinzel, P., Dorotovič, I., Rutten, R.J. (eds.) *The Physics of Chromospheric Plasmas*, *ASP Conf. Ser.* **368**, 217.

Wallace, L., Hinkle, K., Livingston, W.: 2007, *An Atlas of the Spectrum of the Solar Photosphere from 13,500 to 33,980 cm*$^{-1}$ *(2942 to 7405* Å). National Solar Observatory.

DOI 10.1007/978-3-319-12123-9_2
Reprinted from *Solar Physics* Journal, DOI 10.1007/s11207-012-0196-1

FISS Observations of Vertical Motion of Plasma in Tiny Pores

K.-S. Cho · S.-C. Bong · J. Chae · Y.-H. Kim · Y.-D. Park · Y. Katsukawa

Received: 20 December 2011 / Accepted: 7 November 2012 / Published online: 15 January 2013
© The Author(s) 2013. This article is published with open access at Springerlink.com

Abstract Pores can be exploited for the understanding of the interaction between small-scale vertical magnetic field and the surrounding convective motions as well as the transport of mechanical energy into the chromosphere along the magnetic field. For better understanding of the physics of pores, we investigate tiny pores in a new emerging active region (AR11117) that were observed on 26 October 2010 by the Solar Optical Telescope (SOT) on board *Hinode* and the Fast Imaging Solar Spectrograph (FISS) of the 1.6 meter New Solar Telescope (NST). The pores are compared with nearby small magnetic concentrations (SMCs), which have similar magnetic flux as the pores but do not appear dark. Magnetic flux density and Doppler velocities in the photosphere are estimated by applying the center-of-gravity method to the *Hinode*/Spectro-Polarimeter data. The line-of-sight motions in the lower chromosphere are determined by applying the bisector method to the wings of the Hα

Initial Results from FISS
Guest Editor: Jongchul Chae

Electronic supplementary material The online version of this article (doi:10.1007/s11207-012-0196-1) contains supplementary material, which is available to authorized users.

K.-S. Cho (✉) · S.-C. Bong · Y.-H. Kim · Y.-D. Park
Korea Astronomy and Space Science Institute, Daejeon 305-348, Korea
e-mail: kscho@kasi.re.kr

K.-S. Cho
NASA Goddard Space Flight Center, Greenbelt, MD 20771, USA

K.-S. Cho
Department of Physics, The Catholic University of America, Washington, DC 20064, USA

J. Chae
Astronomy Program, Department of Physics and Astronomy, Seoul National University, Seoul 151-747, Korea
e-mail: chae@astro.snu.ac.kr

Y. Katsukawa
National Astronomical Observatory of Japan, 2-21-1 Osawa, Mitaka, Tokyo 181-8588, Japan
e-mail: yukio.katsukawa@nao.ac.jp

and the Ca II 8542 Å line simultaneously taken by the FISS. The coordinated observation reveals that the pores are filled with plasma which moves down slowly and are surrounded by stronger downflow in the photosphere. In the lower chromosphere, we found that the plasma flows upwards inside the pores while the plasma in the SMCs is always moving down. Our inspection of the Ca II 8542 Å line from the wing to the core shows that the upflow in the pores slows down with height and turns into downflow in the upper chromosphere while the downflow in the SMCs gains its speed. Our results are in agreement with the numerical studies which suggest that rapid cooling of the interior of the pores drives a strong downflow, which collides with the dense lower layer below and rebounds into an upflow.

Keywords Chromosphere · Imaging spectroscopy · Photosphere · Pore · Surface magnetic field

1. Introduction

While sunspots can grow to tens of thousands of kilometers, pores are patches of strong magnetic field only about a few Mm wide at the photosphere without penumbral structures. Their simple magnetic field configuration allows investigation of the dynamic interaction between solar magnetic elements and granular convection, and of the transport of energy by flows or waves along strong magnetic field to the chromosphere and corona.

The photospheric appearance of pores has been investigated using ground- and space-based observations. High resolution observations with ground-based telescopes reveal that pores

i) are magnetized areas whose (magnetic) radius is larger than its continuum radius (Keppens and Martinez Pillet, 1996),
ii) are structured with fine structures (*e.g.* an umbral dot and a light bridge),
iii) develop/decay by the accumulation and dispersion of surrounding magnetic knots (*e.g.* Parker, 1979; Hirzberger, 2003), and
iv) are surrounded by downflow channels (Keil *et al.*, 1999).

Recently, *Hinode*/SOT observation by Cho *et al.* (2010) confirmed that the plasma in tiny pores ($R < 2''$) is always moving down, with speeds between 100 and 500 m s^{-1}. They found that the darkness increases as the surrounding plasma flows toward them and decreases as the plasma moves away from them. The plasma in the tiny pores in the photosphere is always moving down and the pores are surrounded by a strong neighboring downward flow (highly red-shifted). There are several numerical simulations of magnetic flux tubes in the solar photosphere. For example, Cameron *et al.* (2007) and Knölker and Schüssler (1988) predicted downflows at the pores. Steiner *et al.* (1998) and Kato *et al.* (2011) carried out numerical simulations of dynamic interaction between magnetic flux tubes and the surrounding gas motion. Especially, Kato *et al.* (2011) reported that the convective downdrafts in the surroundings of a magnetic element in the photosphere leads to a slow and upwardly propagating magnetoacoustic wave, which develops into a shock wave in the chromospheric layers of the magnetic flux tube. Meanwhile, very few chromospheric observations above the pores and small magnetic concentrations (SMCs) have been made, since observations of chromospheric manifestations of pores/SMCs are difficult. This is a major impediment to the understanding of the physics of pores.

The Fast Imaging Solar Spectrograph (FISS; Ahn *et al.*, 2008; Chae *et al.*, 2012) in the Coudé room of the 1.6 m New Solar Telescope (NST; Goode *et al.* 2003, 2010) at Big Bear Solar Observatory is a unique system that can perform fast-scan spectroscopy of extended areas. Since the NST/FISS is suited to improve the current understanding of the dynamics above the pores/SMCs, we have undertaken a coordinated observation with the Solar Optical Telescope (SOT; Tsuneta *et al.*, 2008), on board the *Hinode* satellite (Kosugi *et al.*, 2007). Our observation target was a new emerging active region (AR11117) on 26 October 2010.

2. Observation and Analysis

A new emerging active region (AR11117) appeared from the east limb on 19 October 2010 with a single stable round sunspot. As the sunspot passed across the disk, a chain of tiny sunspots at the south-east of the main sunspot emerged and became much larger and stretched. When the sunspots were born, they had both a preceding sunspot to the west and a following sunspot groups to the east. Figure 1(a) shows the G-band image of the region taken with *Hinode*/SOT. Contours in Figure 1(b) indicate the line-of-sight magnetic fields observed with Helioseismic and Magnetic Imager (HMI; Schou *et al.*, 2012) onboard the *Solar Dynamic Observatory* (SDO; Pesnell, Thompson, and Chamberlin, 2011) superposed on the continuum image from SDO/HMI. It can be seen that the preceding sunspot had a negative polarity and the following sunspot groups had a positive polarity. The pores and SMCs had a negative magnetic field, and they were located between the preceding sunspot and the following sunspot groups as marked by the small box in Figure 1(b).

We have carried out a joint observation of AR11117 with NST/FISS and *Hinode*/SOT from 25 October to 29 October 2010. The photospheric observation of the pores was mainly done by *Hinode*/SOT. G-band and Ca II H-line (3968 Å) images were obtained with the Broad-band Filter Imager (BFI) of *Hinode*/SOT with a pixel size of $0''.16$ and 30 s time cadence for a field of view (FOV) of $75''\times 75''$. Basic data processing was performed by using the SolarSoft package (http://www.lmsal.com/solarsoft). G-band images were aligned to track the pores and SMCs by using the local correlation tracking technique, and the Ca II images were shifted by the same amount to align them with simultaneous G-band images by assuming the same pointing of the G-band and Ca II images by the BFI. The Spectro-Polarimeter (SP) of *Hinode*/SOT recorded the magnetically sensitive Fe I 6301.5/6302.5 lines with a spectral resolution of ≈ 21 mÅ. The SP mapped the region with a normal mode raster scan of 512 steps and time cadence of ≈ 50 min. The effective pixel size was $0''.16$ along the slit and $0''.15$ across it. The field of view was about $75'' \times 81''$. Dark, flat, and wavelength corrections on the SP data were performed by using the SolarSoft package. We have derived line-of-sight magnetic field strengths and line-of-sight velocities from the SP data by using the center-of-gravity (COG) method of Rees and Semel (1979). The SP scans were aligned with BFI images by over-plotting SP continuum data on G-band images.

For the observation of the lower chromosphere, we ran NST/FISS during 17 – 19 UT from 25 October to 29 October. Our observation target was the pores near AR11117. The seeing at the Big Bear Observatory was fairly good on 26 October so that we selected FISS data taken on that day for the present study. FISS scanned the region including the pores located in the small box of Figure 1(a). The FISS obtained raster-scanned spectral images in two spectral lines, Hα and Ca II 8542 Å, simultaneously. The FISS mapped the region with the default mode in which the scanner moves one step, waits for each camera to take one frame, and then moves to the next step. Each step took 0.15 s and the slit was moved over 400 steps. The raster cadence was about 63 s. The effective pixel size was $0''.16$ along and across the slit. The number of pixels along the slit was about 256 so that the field of view was about $40''$ in the slit direction. The scan width was about $64''$. The spectral resolution

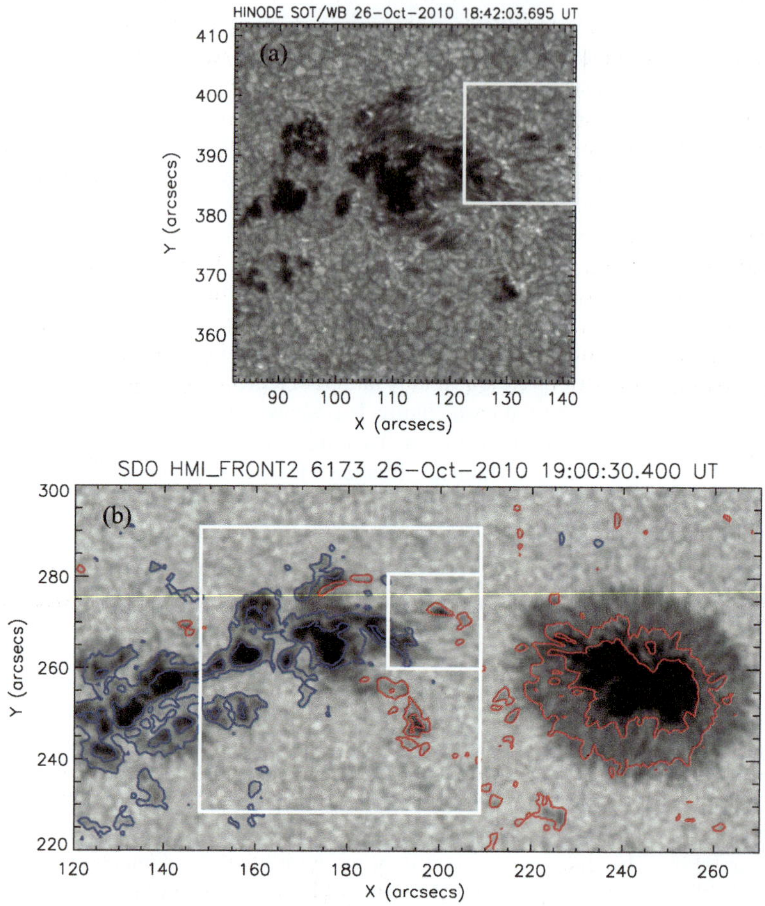

Figure 1 (a) G-band image of a new emerging active region (AR11117) taken with *Hinode*/SOT at 18:42 UT and (b) SDO/HMI continuum image taken at 19:00 UT on 26 October 2010. AR11117 was located near the disk center (N22W16). The large box in the lower panel denotes the field of view of *Hinode*/SOT, and the small box in both panels denotes the region of our interest. The contours in panel (b) indicate the line-of-sight magnetic fields obtained with SDO/HMI. Blue and red contours denote the boundary of positive and negative magnetic fields with their strength levels of 400 G and 1000 G. There was a data gap in SDO observation between 16:13 UT and 19:00 UT.

of FISS is approximately 19 mÅ and 21 mÅ in Hα and Ca II 8542 Å, respectively. Details of basic processing including data acquisition, flat fielding, correction, and compression are described by Chae *et al.* (2012). For wavelength calibration of the spectral lines, we used the two telluric lines of water vapor in the vicinity of each line (6561.097/6565.545 Å for Hα, 8540.817/8546.222 Å for Ca II).

As shown in Figure 2, chromospheric features are dominant in the line-center images, while the outer wings gradually sample the photosphere. This means that the images at different wavelengths from the wing to the center of the Hα and Ca II 8542 Å lines show different features from the photosphere to the chromosphere. It looks likely that the pores and SMCs were covered by fibrils. Both lines showed fibrils with bright feet. Generally, darker Hα fibrils correspond to higher density, and Ca II brightness is more related to temperature (Cauzzi *et al.*, 2009).

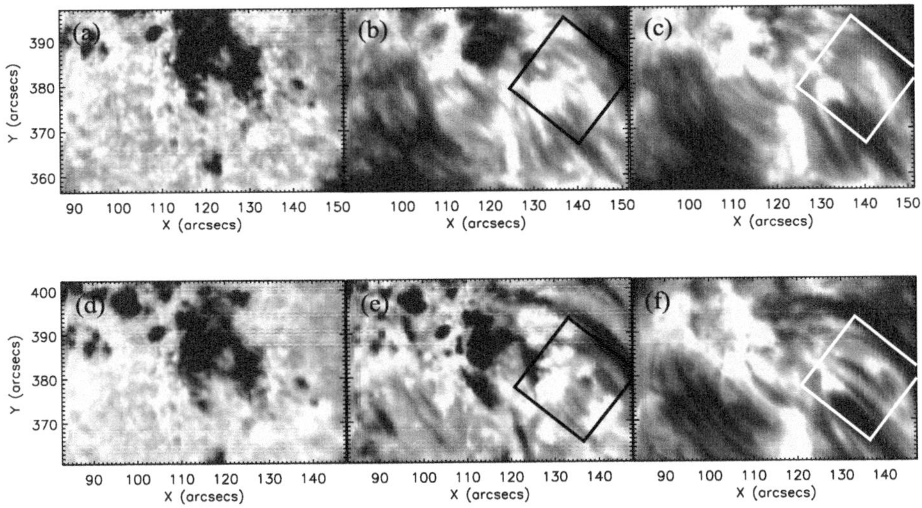

Figure 2 FISS images obtained at 18:41 UT on 26 October 2010 at (a) the wing (−2.5 Å), (b) near the core (−0.2 Å), and (c) the center of the Ca II 8542 Å line, and (d) the wing (−2.0 Å), (e) near the core (−0.7 Å), and (f) the center of the Hα line, respectively. The field of view is about $64'' \times 40''$ and the box in each panel denotes the ROI.

3. Results

3.1. General Characteristics of the Pore and SMC

Figure 3 shows the observation by *Hinode*/SOT for the ROI which is marked with a small box in Figure 1(a). A part of the penumbra of the following sunspot with positive polarity in AR11117 can be seen. The pores and SMCs were located at the network near the penumbra of the following sunspot in AR11117 at the disk center. As shown in Figure 3(b), we can identify four strong magnetic field areas with negative polarity as denoted by white plus symbols. As shown in the G-band image of Figure 1(a), two larger magnetic regions correspond to dark features (pores; P1 and P2), while the other two regions (SMCs; S1 and S2) with strong magnetic flux do not show any dark features.

Why do not the SMCs appear dark in the G-band image even though they are similar to the pores in magnetic flux density? A standard explanation by Spruit and Zwaan (1981) is that pores as well as sunspots are dark by inhibition of convection, while smaller flux tubes such as SMCs may not necessarily be dark because of radiative energy input from their hot walls into the evacuated interior. For a similar reason, the filling fraction of magnetic field in a single pixel was proposed as an alternative explanation of the difference between pores and SMCs (Knölker and Schüssler, 1988). Morinaga *et al.* (2008) reported that the filling factor of pores (0.8 – 0.9) is bigger than that of G-band bright points (0.6). We have estimated the mean filling factor values inside our pores by applying the Milne–Eddington (ME) inversion method (Skumanichi and Lites, 1987) to our SP data. As a result, we found that the filling factor in pores (0.85 for P1 and 0.84 for P2) is a little bigger than that of SMCs (0.81 for SMC1 and 0.77 for SMC2), being in agreement with these studies.

The line-of-sight velocity in the photosphere deduced from SP data shown in Figure 3 suggests that the tiny pores are surrounded by strong downflows, and are filled with weak downflows inside them. This is consistent with the results of Cho *et al.* (2010), who suggested that slow downflows inside pores and fast downflows outside may be associated

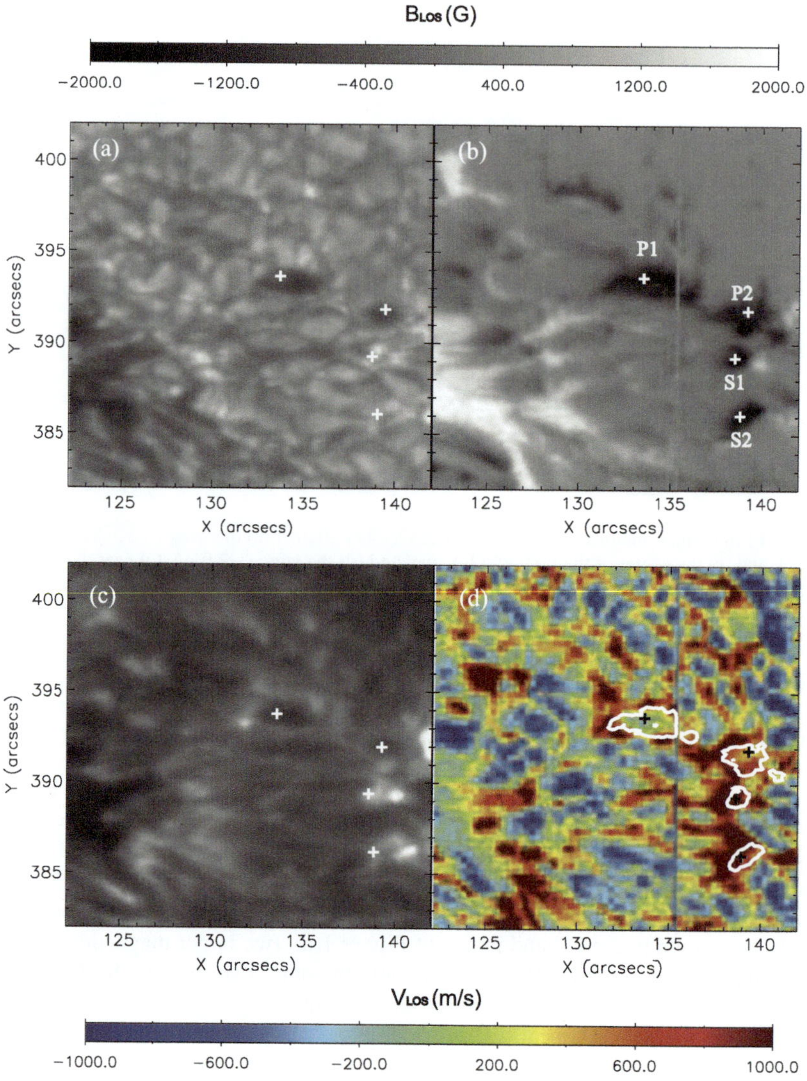

Figure 3 *Hinode*/SOT observations of (a) BFI G-band, (b) SP line-of-sight magnetic field, (c) BFI Ca II H-line core intensity, and (d) SP Doppler map (red color means downdraft) of ROI specified by the small box in Figure 2. Plus symbols represent the pores and SMCs studied here. The white contour in the Doppler map (d) marks the boundaries of strong magnetic field patches (1500 G) in panel (b).

with the formation and maintenance of pores. The downflow channels around the pores and SMCs were well known from previous observations (*e.g.* Keil *et al.*, 1999; Nagata *et al.*, 2008; Morinaga *et al.*, 2008) and from numerical simulations (Cameron *et al.*, 2007; Spruit, 1979).

The pores considered in this study are fairly small (diameter $< 2''$). The time evolution of tiny pores may be different from ordinary pores so that it might be interesting to investigate how these intense magnetic structures evolve during our observation. Unfortunately, there were no SDO/HMI data and *Hinode*/SOT Narrow-band Filter Imager data during our obser-

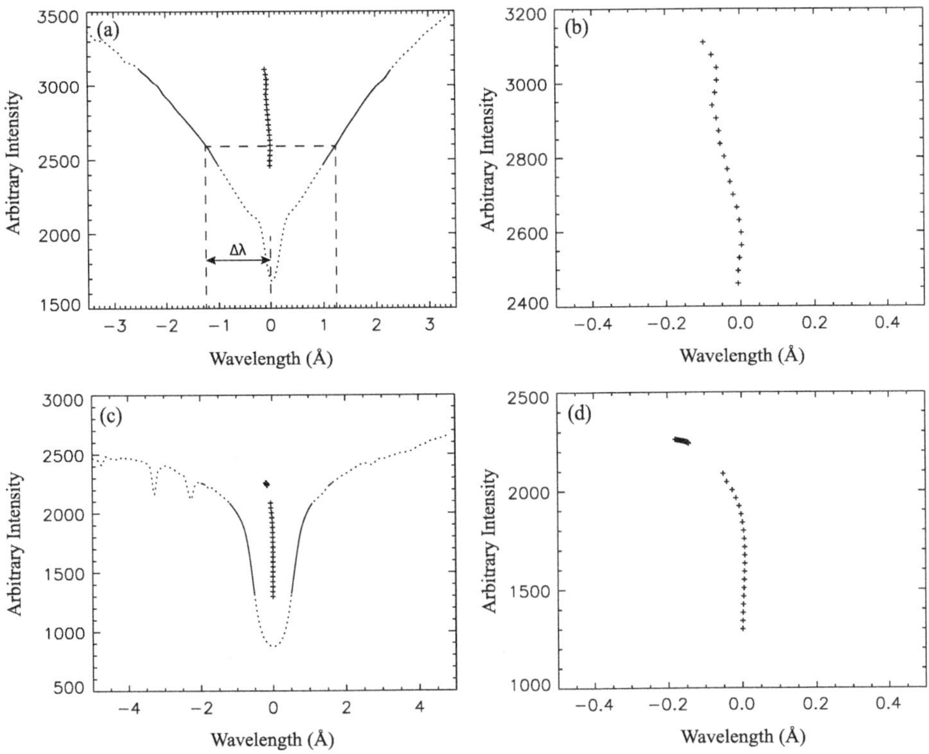

Figure 4 (a) FISS Ca II 8542 Å and (c) Hα line profiles averaged over the FISS FOV. Plus symbols denote the bisector in the wavelength ranges (solid lines) that are used for estimation of the line-of-sight velocity. Enlarged views of the bisector are given in (b) Ca II 8542 Å and (d) Hα. The intensity level for estimation of the bisector is determined at the wavelength offset ($\Delta\lambda$) as shown in panel (a).

vation period, so we could not directly inspect the evolution of the magnetic field. However, we examined the evolution of the pores by using *Hinode*/SOT-BFI G-band observation. The G-band movie (electronic supplementary material) shows that the pores underwent temporal change in intensity and size, but they did not significantly grow or decay during the observation.

3.2. Vertical Motion in the Lower Chromosphere

The Hα line is formed either in chromospheric fibrils or in deep photosphere whereas the Ca II 8542 Å line samples the layers in between the fibrils and the photosphere (Cauzzi *et al.* 2008, 2009). Leenaarts *et al.* (2006) reported that the intensity in the far wings of the Ca II 8542 Å line is highly sensitive to the presence of magnetic structures. In this study, we determined the line-of-sight velocity by applying the bisector measurement to the line profiles of Hα and Ca II 8542 Å acquired by FISS. The line profiles of Ca II 8542 Å (Figure 4(a)) and Hα (Figure 4(b)) are spatially averaged over the FISS FOV as shown in Figure 2. To get the line profile, we applied wavelength correction using the telluric lines as described in the previous section, and then smoothed the spectrum with a 0.1 Å window. The bisector was applied to the wing spectra of the Ca II 8542 Å and Hα lines, respectively. The dotted lines in Figures 4(b) and 4(d) exhibit curved inverse-C-shaped bisectors, which are

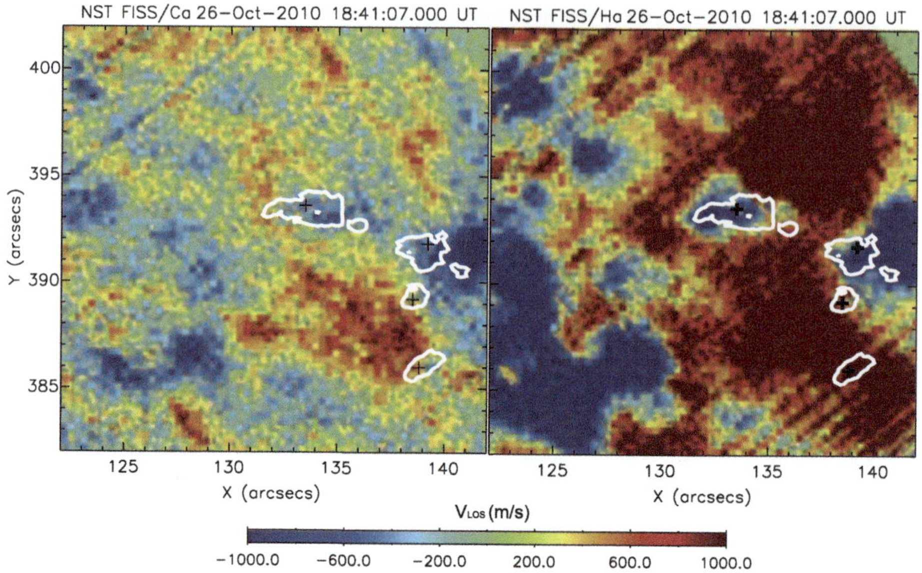

Figure 5 FISS Doppler maps (bisector shift from the nominal line-center wavelength) in the wings of Ca II 8542 Å (left panel, $\Delta\lambda = 2.3$ Å) and Hα (right panel, $\Delta\lambda = 2$ Å), which are taken at 18:41 UT on 26 October 2010. Plus symbols and white contours are the same as those in Figure 3.

in contrast to the regular C-shape of photospheric spectral line bisectors. On the origin of the inverse-C-shape, Uitenbroek (2006) proposed a possibility that it might be the result of upwardly traveling acoustic waves. It should be noted that the inverse-C-shape which was investigated by Uitenbroek (2006) was from the line core of the Ca II 8542 Å line, while our inverse-C-shape is from the wing of the Ca II 8542 Å line.

Since the intensity level in the line spectra reflects the formation height as shown in Figure 2 (Cauzzi *et al.* 2008, 2009), the bisectors measured from the wings of the line represent the Doppler shift at a lower altitude of the atmosphere. In this study, we considered the wavelength ranges of $\Delta\lambda = 1.0-2.5$ Å in Ca II 8542 Å, and $\Delta\lambda = 1.9$ Å and $\Delta\lambda = 0.5-1.2$ Å in Hα, respectively.

Figure 5 presents line-of-sight (LOS) velocity maps constructed by using the bisectors at the wings of Ca II 8542 Å (left) and Hα (right). To obtain the Hα Doppler map, we applied the bisector method to the spectral intensity level near $\Delta\lambda = 2$ Å at each pixel and subtracted the reference velocity which is from the spectrum averaged over the whole scan region and added the disk-center average photosphere value (-120 m s^{-1}) given by Martinez Pillet, Lites, and Skumanich (1997). Since one of the purposes of this study is to inspect the LOS velocity change with height (wavelength) from the photosphere to the chromosphere by comparing *Hinode*/SP and NST FISS observations, we need a reference LOS velocity. We can use the spatially averaged LOS velocity in the chromosphere as the reference velocity, but the averaged velocity in the chromosphere may change with time. It is well known that the granule is on average blue-shifted, so we adopted the disk-center average photosphere value (-120 m s^{-1}) given by Martinez Pillet, Lites, and Skumanich (1997). The Ca II 8542 Å Doppler map was similarly obtained by using the off-band intensities near $\Delta\lambda = 2.3$ Å. Both maps represent the line-of-sight velocity at a lower atmosphere above the photosphere and show similar velocity distribution in general.

As shown in Figure 5, the penumbra shows the blueshift (upflows), while most of the quiet region has redshift (downflows). Pores and SMCs are located at the boundary of the redshift cells. The white contours in Figure 5 showing the level of 1500 gauss (G) delineate the boundaries of strong magnetic patches obtained by *Hinode*/SP observation, of which spatial resolution is about $0''.2$. The effective spatial resolution of the FISS map is about $1''$ (Chae *et al.*, 2012), which is similar to the seeing limit at BBSO; therefore the accuracy of the alignment between the NST/FISS and *Hinode*/SOT is similar to the resolution of the FISS map. As compared with the *Hinode*/SP Doppler map in Figure 3(d), Figure 5 shows structures larger in scale and therefore the characteristic scales of the velocity field look different. This may be because granule-size photospheric features seen by *Hinode*/SP disappear in the lower chromosphere, and/or the spatial resolution or sensitivity of NST/FISS is lower than that of *Hinode*/SP. In both images, we found upflows on the pores and downflows at the SMCs. The upflow areas at the location of the pores in Figure 5 look larger than the photospheric size of pores in the G-band image (Figure 3(a)) and they are slightly shifted in position relative to the pores. It may be related to the expansion of the flux tube, or swaying of it by the ambient horizontal flows in the upper atmosphere.

We inspected the line-of-sight velocity change across the wavelength of the line. As a result, we found that the velocity changes with wavelength in Ca II 8542 Å but this is not the case in Hα. This is because the Hα line is formed either in chromospheric fibrils or in the deep photosphere whereas the Ca II 8542 Å line samples the layer between them (Cauzzi *et al.* 2008, 2009). We speculate that the velocity changes with wavelength seen in Ca II 8542 Å may reflect an intrinsic property of the vertical flows above the pores/SMCs in the lower chromosphere.

Since we are interested in changes in the flow motion with height above the pores in the lower chromosphere, we investigated the Ca II 8542 Å line in detail. Figure 6 shows the line-of-sight velocity changes with wavelength from core ($\Delta\lambda = 1.25$ Å) to the wing ($\Delta\lambda = 2.35$ Å) of Ca II 8542 Å line. Unlike Figure 5, we subtracted the Doppler velocity averaged over weakly magnetized regions in ROI, since our reference velocity set to the quiet photosphere value at disk center given by Martinez Pillet, Lites, and Skumanich (1997) was not convenient. As a result, the Doppler velocity map looks a little different, and the blueshift on the pores is more prominent in the line-wing image taken at $\Delta\lambda = 2.35$ Å as shown in Figure 6(d). We found that redshift is dominant at the line core (Figure 6(a)), while blueshift is dominant at the wing (Figure 6(d)) in general. The red-shifted cells in the penumbra and near the pore and SMC regions reduce their size as one moves to the far-wing wavelengths. If we accept that the bisector reflects the height variation of the Doppler shift, the downflow in the pores turn to the upflow, and the downflow in the SMCs gets slower as one moves from the upper to the lower chromosphere.

Figure 7 presents the line-of-sight velocity change in the pores and SMCs with wavelength. We measured the mean velocities in the circles in Figure 6 at the pores and SMCs as well as in the weak (< 100 G), intermediate, and strong (> 1200 G) magnetic field regions in the ROI. As shown in Figure 7, the velocity in pore P1 (P2) changes from -700 m s^{-1} (-800 m s^{-1}) at $\Delta\lambda = 2.3$ Å to 600 m s^{-1} (5 m s^{-1}) at $\Delta\lambda = 1.2$ Å, namely, increasing with height. The velocity in the SMCs also shows similar change with height but in the opposite sign (red-shifted values). It is also found that the flow velocity in the strong magnetic field regions changes from blueshift to redshift while the flow velocity in the weak and intermediate field regions does not change significantly with height.

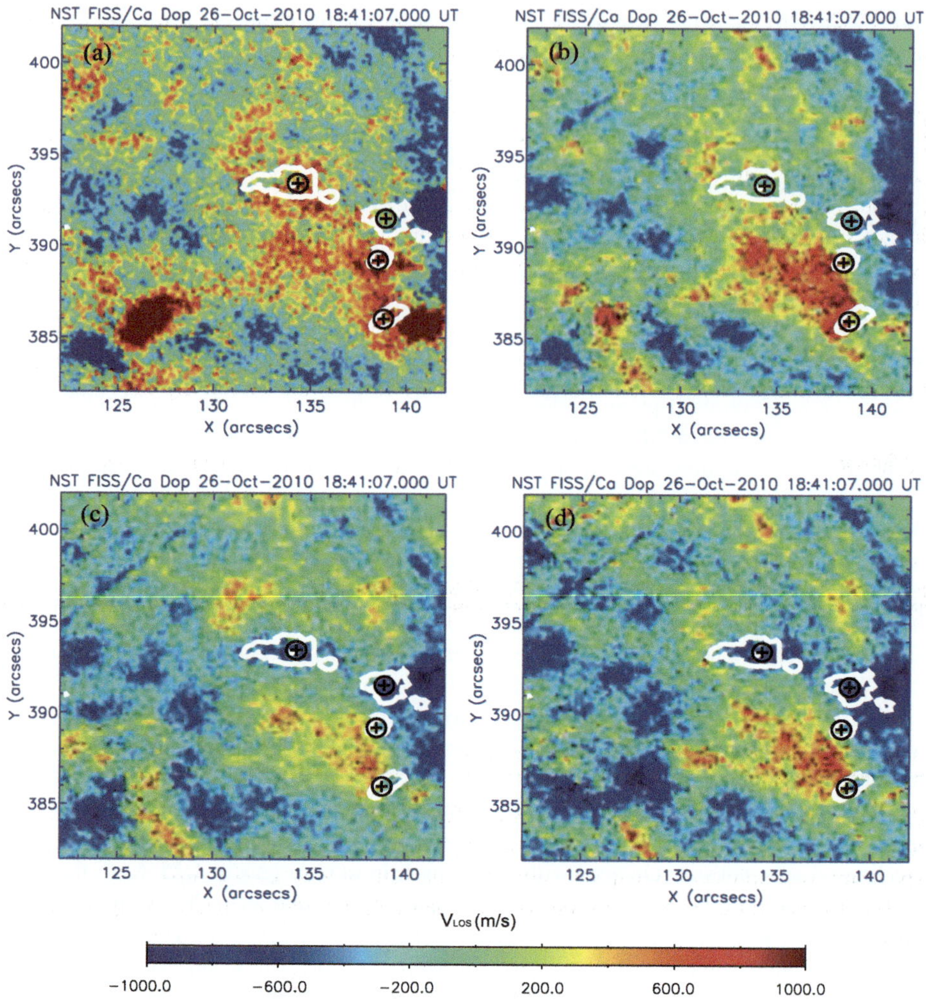

Figure 6 Ca II line-of-sight velocity maps at 18:41 UT on 26 October 2010 at the wavelengths of (a) $\Delta\lambda = 1.25$ Å, (b) $\Delta\lambda = 1.55$ Å, (c) $\Delta\lambda = 1.95$ Å, and (d) $\Delta\lambda = 2.35$ Å. The fringes in panels (c) and (d) are due to data contamination by dust in the FISS observation.

3.3. Evolution of the Vertical Motion

We inspected temporal variations of LOS velocities in the pores and SMCs during the FISS observation period (26 October 2010, 18:30 – 18:57 UT). Figure 8 shows the time variations of the LOS velocities above pore P1 that were estimated at different wavelengths from the line wing to the core. Note that the wavelength-dependence of the LOS velocity as shown in Figure 7 was maintained throughout the observing time; for example, LOS velocities at $\Delta\lambda = 2.35$ Å remained significantly negative and those at $\Delta\lambda = 1.25$ Å remained significantly positive. The LOS velocities in fact changed with time, either slowly varying or oscillating, but such changes were not drastic enough to alter the pattern described above. We obtained similar results for the other pore (P2) and the two SMCs. This characteristic of velocity field in the low chromosphere of the pores and SMCs we found is similar to that of

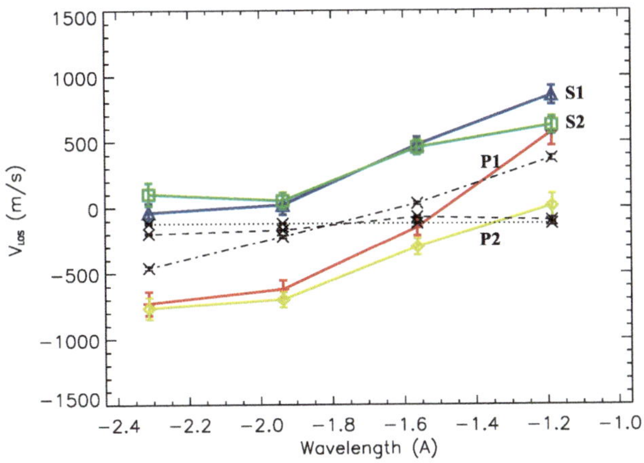

Figure 7 Variation of the line-of-sight velocity in the pores (P1, P2) and SMCs (S1, S2) at various wavelengths from the wing to the core in the Ca II line. Black lines denote the distribution at the strong (dash dotted), intermediate (dashed), and weak (dotted) magnetic field regions. The uncertainty in the mean velocity was estimated from the velocity fluctuation inside the averaged region. The estimated uncertainties are $\approx 100\ \mathrm{m\,s^{-1}}$ for the pores and SMCs, and $\approx 10\ \mathrm{m\,s^{-1}}$ for the strong, intermediate, and weak magnetic regions.

Figure 8 Time variation of the line-of-sight velocity in pore P1 at the wavelengths of $\Delta\lambda = 1.25$ Å (thin gray), $\Delta\lambda = 1.55$ Å (thick gray), $\Delta\lambda = 1.95$ Å (thin black), and $\Delta\lambda = 2.35$ Å (thick black) in the Ca II line.

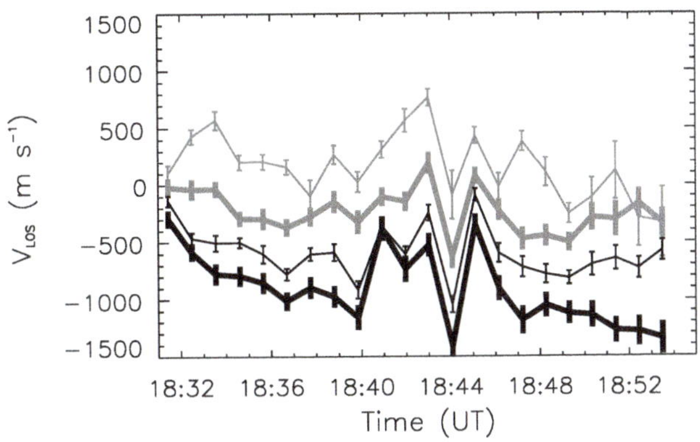

the photospheric velocity of pores as was reported by Hirzberger (2003) and Sobotka *et al.* (2012).

We investigated a dynamic interaction between the magnetic flux tubes (pores and SMCs) and the photospheric gas motion by using the G-band data taken with *Hinode*/SOT during our observation. For horizontal gas motion in the photosphere, we first aligned G-band images and applied subsonic-filtering to reduce solar oscillation effect and noise. We then applied the nonlinear affine velocity estimator (NAVE) to the successive G-band images and got a first-order surface velocity field including the first derivatives with respect to the horizontal coordinates as described by Chae and Sakurai (2008). The divergence of the velocity was obtained from those first derivatives given by the NAVE technique. We tracked the position of the pores and SMCs using NAVE, as seen in the G-band movie (electronic supplementary material). Finally, we measured the averaged horizontal velocity and divergence within the radius of 0.5″ around the position of the pores and SMCs. The transverse acceleration was derived from the running difference of the averaged horizontal velocity. The LOS

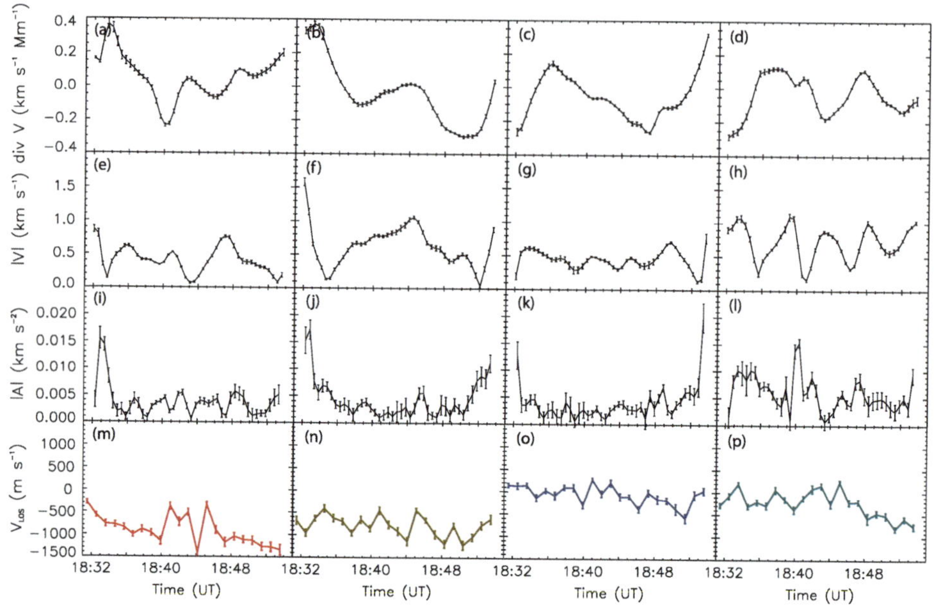

Figure 9 Time variations of (a – d) the divergence of velocity, (e – h) transverse velocity, and (i – l) acceleration at the pores and SMCs, which are estimated from G-band images by applying the NAVE technique. FISS LOS velocities above pores P1 (m), P2 (n), and SMC1 (o) and SMC2 (p) are estimated by applying the bisector method to the wing ($\Delta\lambda = 2.35$ Å) spectra of the Ca II 8542 Å line.

velocity in the chromosphere obtained by FISS was similarly averaged inside the pores and SMCs. Figure 9 shows a comparison between the temporal variations of the photospheric gas motion (divergence, transverse velocity, and acceleration) inside the pores and SMCs, and the chromospheric LOS velocity on the pores and SMCs. We found that the shape and brightness of the pores and SMCs observed in the G-band change with time due to their interaction with granular motion, but there is no clear relationship between the physical parameters representing the photospheric gas motion and the chromospheric LOS velocity. No apparent difference in the LOS velocity between the pores and SMCs was found.

The time evolution of pores and SMCs and their interaction with the surrounding gas motions can be thoroughly studied when the structures are stably imaged with high spatial resolution and sensitivity. Unfortunately, the performance of currently working low-order adaptive optics of NST/FISS is not satisfactory in this regard.

4. Discussion and Conclusion

We have derived the physical characteristics of small pores based on the simultaneous observation of *Hinode*/SOT and NST/FISS. To our knowledge, this is the first detailed report on the connection of LOS plasma flows in tiny pores between the photosphere and the lower chromosphere through the determination of the chromospheric LOS velocities in pores and SMCs simultaneously, as well as their photospheric LOS velocities.

Hinode/SOT-SP observation elucidates that the plasma in the pores is always moving down on the photosphere, while the FISS observation of the wing spectra of Ca II 8542 Å and Hα detected upflows at the pore in the low chromosphere. Figure 10 illustrates a schematic

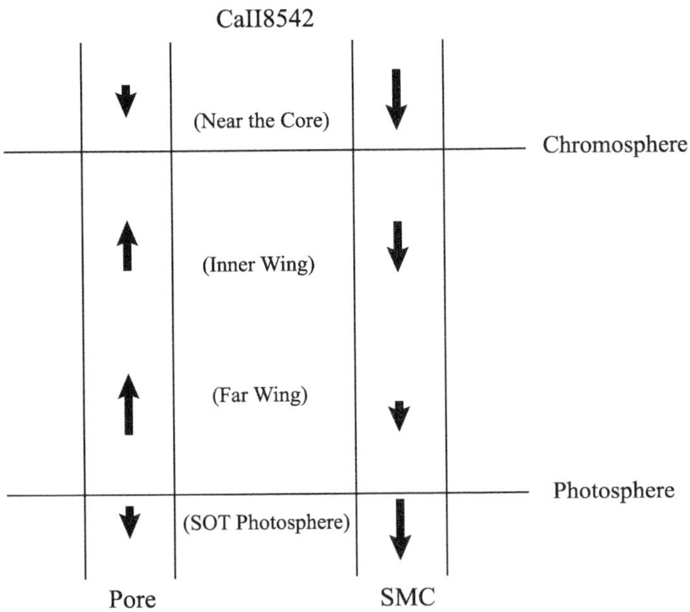

Figure 10 Vertical motion of plasma in the pore and SMC in the photosphere and chromosphere.

picture about the height variation of vertical motion inside pores and in SMCs, respectively. The inspection of the Ca II 8542 Å line from the wing to the core by applying the bisector method revealed that the upflow velocity in the pores decreases with height and turns into downward in the upper chromosphere, while downflows in the SMCs gain their speed in the lower chromosphere. Those lasted at least 20 min during the observational period (Figure 8).

Our observations on the dynamic property of flux tubes could be understood by considering a dynamic interaction between solar magnetic elements and granular convection simulated by Steiner *et al.* (1998) and Kato *et al.* (2011). According to their two-dimensional MHD simulation, the interior of a pore cools rapidly and induces strong downflow around the pore owing to radiative losses and suppression of energy transport by convection in the pore. Their simulation predicts that the downflow turns into upflow after interacting with a dense and deep layer under the pore. This theoretical description is well consistent with our observations of LOS velocities in the photosphere and low chromosphere. It is likely that only strong downflow can excite an upward propagating shock wave as predicted by Grossmann-Doerth, Schüssler, and Steiner (1998), and it is expected that the upward shock is continuously observed in the upper photosphere and lower chromosphere.

In SMCs, we detected downflows, but no upflows at all. This might be explained by the absence of localized rapid cooling in SMCs because of the heat coming from the surroundings. In other words, the existence of downflows only suggests that a more or less continuous process may be more important in SMCs than a localized, impulsive cooling process unlike pores. If the downflow at the photosphere is caused by the convective collapse, the time scale of the event is several minutes (*e.g.* Narayan, 2011). Persistent downflows over 20 min might represent the spatially averaged property of multiple flux tubes which are subject to convective instability.

In summary, we observed upward-directed dynamic motions at the tiny pores in the lower chromosphere by inspecting the wing spectra of chromospheric lines, which might be related to upwardly propagating magnetoacoustic waves predicted by Steiner *et al.* (1998) and Kato *et al.* (2011).

We are expecting that the high-order adaptive optics system for NST will provide stable observations with high spatial/spectral/temporal (10 s) resolutions, which will contribute to understanding the nature of the upward motion in the pores and SMCs, as well as their interaction with ambient plasma.

Acknowledgements We are grateful to the referee for helpful and constructive comments. This work was supported by the "Development of Korea Space Weather Center" of KASI and the KASI basic research funds and by the Korea Research Foundation Grant funded by the Korean Government (KRF-2008-220-C00022, 2011-0028102). One of the authors (K.-S.C.) is very thankful to R.J. Rutten for his valuable comments and to S. Tsuneta for supporting of collaborative visiting in NAOJ, where the part of present work has been carried out. *Hinode* is a Japanese mission developed and launched by ISAS/JAXA, with NAOJ as domestic partner and NASA and STFC (UK) as international partners. It is operated by these agencies in cooperation with ESA and NSC (Norway).

Open Access This article is distributed under the terms of the Creative Commons Attribution License which permits any use, distribution, and reproduction in any medium, provided the original author(s) and the source are credited.

References

Ahn, K., Chae, J., Park, H.M., Nah, J., Park, Y.-D., Jang, B.H., Moon, Y.-J.: 2008, *J. Korean Astron. Soc.* **41**, 39.
Cameron, R., Schüssler, M., Vögler, A., Zakharov, V.: 2007, *Astron. Astrophys.* **474**, 261.
Cauzzi, G., Reardon, K., Uitenbroek, H., Cavallini, F., Falchi, A., Falciani, R., Janssen, K., Rimmele, T., Vecchio, A., Wöger, F.: 2008, *Astron. Astrophys.* **480**, 515.
Cauzzi, G., Reardon, K., Rutten, R.J., Tritschler, A., Uitenbroek, H.: 2009, *Astron. Astrophys.* **503**, 577.
Chae, J., Sakurai, T.: 2008, *Astrophys. J.* **689**, 593.
Chae, J., Park, H.M., Ahn, K., Yang, H., Park, Y.-D., Nah, J., Jang, B.H., Cho, K.-S., Cao, W., Goode, P.R.: 2012, *Solar Phys.* doi:10.1007/s11207-012-0147-x.
Cho, K.-S., Bong, S.-C., Chae, J., Kim, Y.-H., Park, Y.-D.: 2010, *Astrophys. J.* **723**, 440.
Goode, P.R., Denker, C.J., Didkovsky, L.I., Uhn, J.R., Wang, H.: 2003, *J. Korean Astron. Soc.* **36**, S125.
Goode, P.R., Coulter, R., Gorceix, N., Yurchyshyn, V., Cao, W.: 2010, *Astron. Nachr.* **331**, 620.
Grossmann-Doerth, U., Schüssler, M., Steiner, O.: 1998, *Astron. Astrophys.* **337**, 928.
Hirzberger, J.: 2003, *Astron. Astrophys.* **405**, 331.
Kato, Y., Steiner, O., Steffen, M., Suematsu, Y.: 2011, *Astrophys. J. Lett.* **730**, L24.
Keppens, R., Martinez Pillet, V.: 1996, *Astron. Astrophys.* **4**, 291.
Keil, S.L., Balasubramaniam, K.S., Smaldone, L.A., Reger, B.: 1999, *Astrophys. J.* **510**, 422.
Knölker, M., Schüssler, M.: 1988, *Astron. Astrophys.* **202**, 275.
Kosugi, T., Matsuzaki, K., Sakao, T., Shimizu, T., Sone, Y., Tachikawa, S., *et al.*: 2007, *Solar Phys.* **243**, 3.
Leenaarts, J., Rutten, R.J., Carlsson, M., Uitenbroek, H.: 2006, *Astron. Astrophys.* **452**, L15.
Martinez Pillet, V., Lites, B.W., Skumanich, A.: 1997, *Astrophys. J.* **474**, 810.
Morinaga, S., Sakurai, T., Ichimoto, K., Yokoyama, T., Shimojo, M., Katsukawa, Y.: 2008, *Astron. Astrophys.* **481**, L29.
Nagata, S., Tsuneta, S., Suematsu, Y., Ichimoto, K., Katsukawa, Y., Shimizu, T., *et al.*: 2008, *Astrophys. J. Lett.* **677**, L145.
Narayan, G.: 2011, *Astron. Astrophys.* **529**, A79.
Parker, E.N.: 1979, *Astrophys. J.* **234**, 333.
Pesnell, W.D., Thompson, B.T., Chamberlin, P.C.: 2011, *Solar Phys.* **275**, 3.
Rees, D.E., Semel, M.D.: 1979, *Astron. Astrophys.* **74**, 1.
Schou, J., Scherrer, P.H., Bush, R.I., Wachter, R., Couvidat, S., Rabello-Soares, M.C., *et al.*: 2012, *Solar Phys.* **275**, 229.
Skumanichi, A., Lites, B.W.: 1987, *Astrophys. J.* **322**, 473.
Sobotka, M., Del Moro, D., Jurčák, J., Berrilli, F.: 2012, *Astron. Astrophys.* **537**, A85.

Spruit, H.C.: 1979, *Solar Phys.* **61**, 363.
Spruit, H.C., Zwaan, C.: 1981, *Solar Phys.* **70**, 207.
Steiner, O., Grossmann-Doerth, U., Knoelker, M., Schuessler, M.: 1998, *Astrophys. J.* **495**, 468.
Tsuneta, S., Ichimoto, K., Katsukawa, Y., Lites, B.W., Matsuzaki, K., Nagata, S., *et al.*: 2008, *Solar Phys.* **249**, 167.
Uitenbroek, K.H.: 2006, *Astrophys. J.* **639**, 516.

DOI 10.1007/978-3-319-12123-9_3
Reprinted from *Solar Physics* Journal, DOI 10.1007/s11207-013-0354-0

Velocities and Temperatures of an Ellerman Bomb and Its Associated Features

Heesu Yang · Jongchul Chae · Eun-Kyung Lim · Hyungmin Park · Kyuhyoun Cho · Ram Ajor Maurya · Donguk Song · Yeon-Han Kim · Philip R. Goode

Received: 28 December 2012 / Accepted: 21 June 2013 / Published online: 26 July 2013
© Springer Science+Business Media Dordrecht 2013

Abstract We investigated the velocity and temperature characteristics of an Ellerman bomb (EB) and its associated features based on observations made with the Fast Imaging Solar Spectrograph (FISS) and a broadband TiO filter of the 1.6 meter New Solar Telescope at Big Bear Solar Observatory. In the TiO images of the photospheric level, we found a granular cell expanding in two opposite directions near the site of the EB. When one end of this granule reached the EB site, the transverse speed of the tip of the expanding granule rapidly decreased and the EB brightened. The wings of the Hα profile of the EB indicated that the EB was blueshifted up to 7 $\mathrm{km\,s^{-1}}$. About 260 s after the EB brightening, a surge was seen in absorption and varied from a blueshift of 20 $\mathrm{km\,s^{-1}}$ to a redshift of 40 $\mathrm{km\,s^{-1}}$ seen in the Hα and Ca II 8542 Å lines. From the Doppler absorption width of the two lines determined by applying the cloud model, we estimated the mean temperature of the surge material to be about 29000 K and the mean speed of nonthermal motion to be about 11 $\mathrm{km\,s^{-1}}$. We discuss the physical implications of our results in terms of magnetic reconnection and processes related to it.

Keywords Active regions, velocity field · Chromosphere, active · Magnetic reconnection, observational signatures · Spectrum, visible · Velocity fields, photosphere

Initial Results from FISS
Guest Editor: Jongchul Chae

H. Yang (✉) · J. Chae · H. Park · K. Cho · R.A. Maurya · D. Song
Astronomy Program, Department of Physics and Astronomy, Seoul National University, Seoul 151-747, Korea
e-mail: yang@astro.snu.ac.kr

J. Chae
e-mail: chae@astro.snu.ac.kr

E.-K. Lim · Y.-H. Kim
Korea Astronomy & Space Science Institute, Daejeon 305-348, Korea

P.R. Goode
Big Bear Solar Observatory, Big Bear City, CA 92314-9672, USA

1. Introduction

An Ellerman bomb (EB) is a small feature in an active region that appears bright at the far wings of the Hα line (Ellerman, 1917). It has a short lifetime of about 10 min, and a size of about 1$''$ (Kurokawa *et al.*, 1982). It frequently occurs in an emerging flux region or near a sunspot. It usually has internal mass motions that often accompany a surge (Georgoulis *et al.*, 2002).

At the site of EBs, two types of internal mass motions were reported so far: downflows in the photosphere and upflows in the low chromosphere. Georgoulis *et al.* (2002) found that 80 % of the EBs have downward flows of about 0.1 to 0.3 km s^{-1} at the photospheric level. In the low chromosphere, Kitai (1983) and Kurokawa *et al.* (1982) detected upflows with a speed of 6 to 8 km s^{-1} from the far blue wings of the Hα line. In addition, transverse motions associated with EBs were observed in the photosphere (Jess *et al.*, 2010; Payne, 1993; Georgoulis *et al.*, 2002).

Some previous studies have reported that EBs were accompanied by surges (Altschuler, Lilliequist, and Nakagawa, 1968; Rust, 1968; Roy, 1973; Dara *et al.*, 1997; Matsumoto *et al.*, 2008; Watanabe *et al.*, 2011; Canfield, 1996). Matsumoto *et al.* (2008) found the ejection of Hα-absorbing material from an EB with a speed of 5 km s^{-1} in the plane of the sky. Watanabe *et al.* (2011) also found similar events in images taken at several wavelengths of the Hα line with a Fabry-Perot interferometer. They examined 11 EBs and found that two of them were associated with surges. One of them initially appeared as an upflow with a line-of-sight (LOS) velocity of about 40 km s^{-1}; this EB later changed to a downflow. They recurred a few times during the observations.

Previous studies suggested that EBs occur in specific field configurations. Roy (1973) and Dara *et al.* (1997) showed that EBs are located near the boundaries of magnetic features. Georgoulis *et al.* (2002) showed that 70 % of the EBs were associated with bald-patch field lines, which represent magnetic dips of the U-loops in the photosphere, and the remaining 30 % of the EBs were associated with the quasi-separatrix layers in apparently monopole regions. This pattern was confirmed by other studies (Pariat, Aulanier, and Schmieder, 2004; Fang *et al.*, 2006; Pariat *et al.*, 2007).

It is now widely accepted that the driving mechanism of EBs is magnetic reconnection in the temperature-minimum region or in the low chromosphere (Pariat, Aulanier, and Schmieder, 2004; Georgoulis *et al.*, 2002; Pariat *et al.*, 2007; Isobe, Tripathi, and Archontis, 2007). Magnetic reconnection could explain the observed mass motions in the photosphere and chromosphere, including surges (Matsumoto *et al.*, 2008), a typical elongated shape of the inverse-Y shape (Watanabe *et al.*, 2011), and the local heating of the EBs with $\Delta T \approx 500 - 2000$ K (Kitai, 1983; Fang *et al.*, 2006; Georgoulis *et al.*, 2002).

In this paper, we investigate the velocities and temperatures of plasmas of an EB and its associated features. We are interested in a variety of processes related to the same EB: the photospheric transverse motion and downflow at the site of the EB, the EB itself as a jet-like feature in the chromospheric level, and an intensely heated surge. The paper proceeds as follows. In Section 2 we introduce the observation and data processing. Section 3 contains the results, particularly on the velocity field inside the EB, the magnetic-field configuration around the EB, and the parameters of the EB-associated surge. Finally, we interpret and discuss our results in Section 4.

2. Observations and Data Processing

We observed a series of EBs near the pores in AR 11271 on 19 August 2011 using the Fast Imaging Solar Spectrograph (FISS) and a TiO broadband filter of the 1.6 m New Solar

Telescope at Big Bear Solar Observatory with the aid of adaptive optics. The FISS is an Echelle spectrograph that produces spatial and spectral information of the Hα and Ca II 8542 Å lines simultaneously, using a fast scan of the slit across the field of view (Chae *et al.*, 2012). The spectral sampling is 19 mÅ in the Hα line and 26 mÅ in the Ca II line, and the spectral coverage is 9.7 Å in the Hα line and 12.9 Å in the Ca II line, which is broad enough to cover the lines. The region was located at 447″ eastward and 146″ northward of the disk center. The field of view (FOV) was 19.2″ in the scan direction and 40″ in the slit direction. The step size is 0.16″, which is the same as the sampling along the slit. The observation lasted for 2 h from 18:54:31 UT to 21:01:10 UT in steps of 20 s.

We performed basic processing of the FISS data – flat fielding, dark subtraction, and distortion correction – following Chae *et al.* (2012). The spatially averaged profiles of the Hα line and Ti II 6559.580 Å photospheric line were used for the wavelength calibration of the Hα band. With this calibration, the spatial average of the Hα Doppler shift becomes zero. In the same way, the spatial averages of the Ca II line and Kr I 8537.93 Å line were used to calibrate the Ca II band. The measured standard error in the profile was 0.15 $\mathrm{km\,s^{-1}}$ for Hα and 0.21 $\mathrm{km\,s^{-1}}$ for the Ca II line in the quiet region.

The region was observed with the TiO broadband filter as well, which is centered at a wavelength of 7057 Å and has a band width of 10 Å (Cao *et al.*, 2010). The best 70 images of the 100 short-exposure (1.2 ms) images in each burst were saved during the observation. After the dark subtraction and flat fielding, we applied the Kiepenheuer-Institut Speckle Interferometry Package (Wöger, von der Lühe, and Reardon, 2008) code to each burst. The FOV of the speckle-reconstructed images was 69.8″×69.8″. Bursts of images were taken repeatedly and the interval between two successive bursts was about 15 s.

Figure 1 shows a few raster images of the observed region constructed from the FISS data at several wavelengths and a speckle-reconstructed image through the TiO broadband filter. The spatial resolution of the FISS images is poorer than that of the TiO image partly because the initial focusing of the FISS was not set to the best focus (Chae *et al.*, 2012) and the atmospheric seeing was not as well corrected by the current adaptive optics as it is now.

The EB of our interest is clearly identified in the 6562.8 Å + 1.4 Å image. It is located at 437″ eastward and 146″ northward of the disk center, about 10″ away from the pores. The angle between the LOS and the radial direction, μ, is about 30°. The surge appeared at the side of the EB and seems to have originated from it (see the Hα and Ca II line center images of Figure 1). In the TiO filter image, we see many tiny bright points corresponding to the footpoints of magnetic flux tubes on the boundaries of granules.

3. Results

We observed three EBs and two surges during 2 h of observations. Our analysis is focused on the second EB and its associated surge because the data quality is better than the others. This event first appeared at 19:26:49 UT, which is taken to be the epoch of time t. Before the EB occurred, an oval-shaped granular cell expanded in the TiO photospheric images. When the lane of the granular cell reached the magnetic neutral line, the expansion stopped. Then the EB brightened and persisted for about 10 min. About 260 s after the EB brightening, we observed that the surge was ejected next to the EB.

The dataset of the first and third EBs is incomplete because of the poor atmospheric conditions. The third EB appeared at $t = 820$ s. We found transverse motions in the TiO images before the brightening. Another surge was observed in the blue wing of the Hα and Ca II lines at $t = 920$ s.

In the next section, we explain the events related to the second EB in detail.

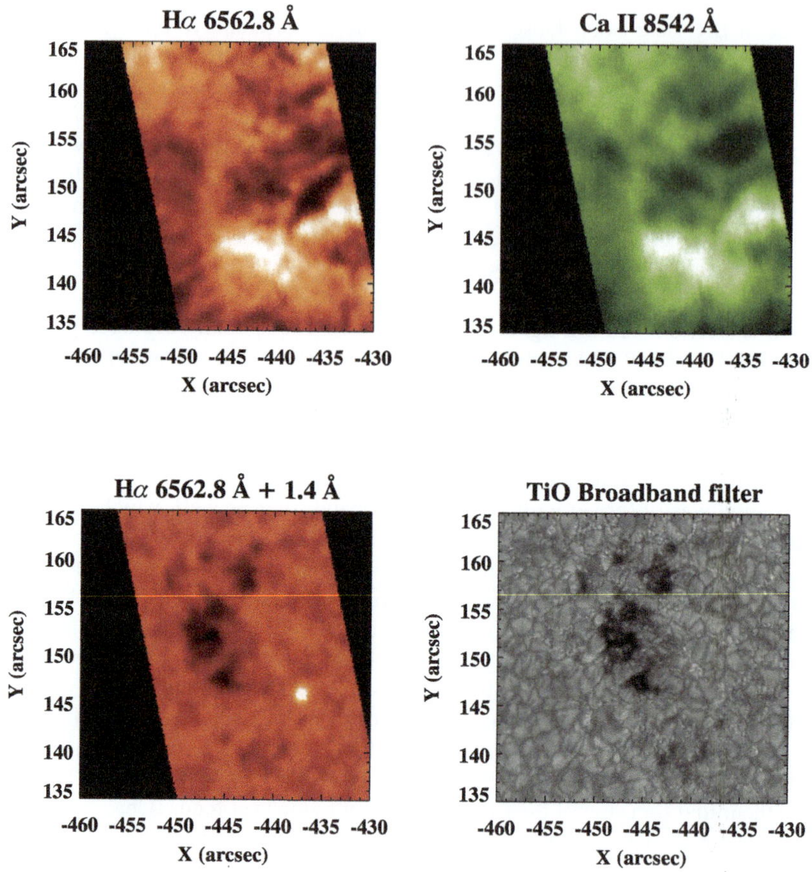

Figure 1 The raster images of the Hα line center (top left), the Ca II 8542 Å line center (top right), and the off-band (bottom left) of the FISS and TiO broadband filter image (bottom right) at 19:32:48 UT, 359 s after the occurrence of the EB.

3.1. Expanding Granular Cell

We found an oval-shaped granular cell expanding in two opposite directions near the site of the EB (see Figure 2). During this expansion period from −800 s to 200 s, new dark lanes emerged inside the cell, which caused the cell to split into four sub-cells. One end of this expanding granule, surrounded by the dark lanes, reached the site where the EB occurred. As this approached the EB region, a small sub-cell appeared at the tip of this lane and then brightened. A set of 45-s cadence magnetograms of the *Helioseismic and Magnetic Imager* (HMI; Schou *et al.*, 2012) onboard the *Solar Dynamics Observatory* (SDO) suggests that the EB was located above the region where the magnetic flux gradient was steep. A weak positive-polarity region approached the EB site together with the expanding granule. At the other side of the granular cell, the steep-gradient magnetic polarity region also moved together with the expanding granule.

We measured the transverse speed, v_{TiO}, of the edge of the expanding granule in the direction toward the EB using the nonlinear affine velocity estimator (NAVE; Chae and Sakurai, 2008). We chose 13 × 13 pixels in the TiO image as the window size, similar to the

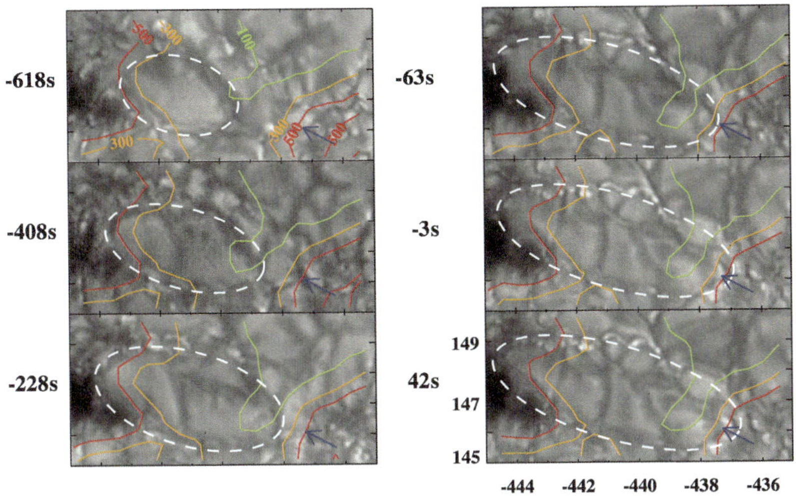

Figure 2 Time series of TiO 7057 Å filter images. The blue arrows indicate the position of the EB. The time after the occurrence of the EB is presented at each image. White dashed ovals indicate the expanding granular cell of interest. The solid curves represent the levels of the vertical magnetic field in the HMI magnetograms.

size of the small sub-cell (0.5″) that appeared at the tip of the lane. As a result, we found that v_{TiO} increased from 1 $\mathrm{km\,s^{-1}}$ up to 5.5 $\mathrm{km\,s^{-1}}$ in 350 s (see Figure 10(d)); this is three times faster than the typical granular flow speed, implying that this motion was not driven by photospheric granulation. After reaching its peak, the speed rapidly decreased to about 1 $\mathrm{km\,s^{-1}}$ within 100 s.

3.2. Photospheric Doppler Velocity

The LOS velocities in the photosphere of the EB were inferred from the photospheric line at 6559.580 Å in the Hα-band spectra by fitting a Gaussian profile to the line. The measured velocities looked somewhat noisy partly because the line is weak and they are subject to solar oscillation. Despite these problems, we found a persistent downflow of about 1 $\mathrm{km\,s^{-1}}$ at the EB location. The downflow started at -100 s, and lasted for about 300 s. This value is larger than the standard error of 0.11 $\mathrm{km\,s^{-1}}$.

3.3. Spectral Characteristics of the EB

Figure 3 shows the spectral characteristics of the EB seen in the Hα and Ca II line profiles. As is well known, the Hα line profile displays a broad emission profile at the wings and a narrow absorption at the center. The intensity monotonically increases in the emission wings as the wavelength approaches the line center and peaks at around ± 1.0 Å. The Ca II line profile of the EB appears in emission at every wavelength when compared with the reference profile of the surrounding area. Unlike the Hα profile, the intensity of the Ca II line decreases as the wavelength approaches the line center from the wings.

The intensity variation over wavelength at the wings reflects the height variation of the source function in the radiative transfer. The wings of the Ca II profile originate from the photosphere where local thermodynamic equilibrium (LTE) holds fairly well. The brightness temperature T_{B} of the photosphere where the EB is observed can be identified with the

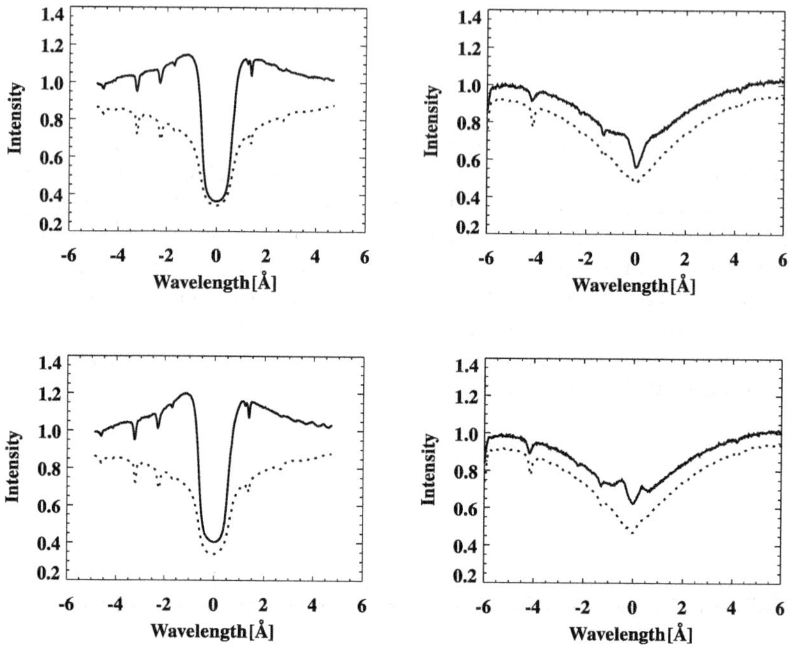

Figure 3 The intensity profiles of the Hα line (left) and the Ca II line (right) of the EB taken at 199 s (top) and 319 s (bottom) after the occurrence of EB. The solid lines represent the profile of the EB, and the dotted lines represent the reference profiles of the surrounding area.

temperature T of the formation region. Figure 3 hence indicates that the photosphere of the EB is characterized by the temperature excess $\Delta T_{\mathrm{B}} \approx 200$ K over the unperturbed one. This increase in the photospheric temperature is consistent with the brightening in the TiO images and with the brightening in the SDO/AIA 1600 Å and 1700 Å images that took place simultaneously with those of the Hα and Ca II lines.

The Hα profile of the EB itself is decomposed into a few components: central absorption, emission in the near wings, and "power-law emission in the far wings" (Kitai, 1983). We inferred the LOS velocity of the plasma responsible for the Hα emission in the near wings, v_{wing}, with the lambdameter method. In this method, v_{wing} is determined from

$$v_{\mathrm{wing}} = c\frac{\lambda_{\mathrm{m}} - \lambda_0}{\lambda_0}, \tag{1}$$

where λ_{m} is a measure of the Doppler shift and λ_0 is the central wavelength of the line. The quantity λ_{m} is defined by

$$C\left(\lambda_{\mathrm{m}} - \frac{\delta\lambda}{2}\right) = C\left(\lambda_{\mathrm{m}} + \frac{\delta\lambda}{2}\right) = C_1, \tag{2}$$

where $C(\lambda)$ is the contrast profile defined in Equation (3). The wavelength difference of the lambdameter, $\delta\lambda$, is set to 2.8 Å, which represents the emission characteristics well, and C_1 is obtained as a by-product of the method. We applied this lambdameter method to the contrast profiles instead of to the spectral profiles because the characteristics of the EB may be better manifested in the contrast profiles.

Figure 4 shows the temporal variation of the contrast profiles of the EB. The emission wings of the EB lasted for about 500 s. During this period, they are blueshifted, and the

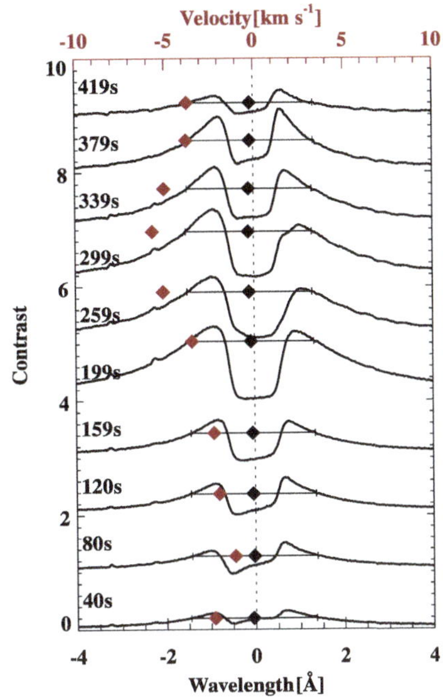

Figure 4 Temporal variation of the Hα contrast profiles of the EB. The contrast profiles are stacked vertically in the order of time. Black horizontal lines and diamond symbols represent the line widths of $\delta\lambda$ (2.8 Å) and the values of λ_m of the emission wings. Red axis and red diamond symbols represent the calculated LOS velocities, v_{wing}.

resulting velocity v_{wing} gradually changed from 1 $\mathrm{km\,s^{-1}}$ to 7 $\mathrm{km\,s^{-1}}$ with an acceleration of 14 $\mathrm{m\,s^{-2}}$.

3.4. Surge

The raster images of the small region presented in Figure 5 show the second and third EBs and the surges associated with three EBs. The second EB is located at the bottom-left of every off-band raster image. After the second EB appeared, it brightened slowly with time and reached a peak at $t = 298$ s, after which it darkened. The surge was visible from about $t = 259$ s in the blue wings of the two lines. Later, the surge disappeared in the blue wings, but emerged in the red wings, indicating the gradual transition of the Doppler shift from blueshift to redshift.

Figure 6 shows the spectral profiles of the Hα (left panel) and Ca II 8542 Å lines (right panel) taken from a spatial point inside the surge at $t = 741$ s. The spectral profiles were normalized with respect to the far wings ($|\lambda| > 2$ Å) of the Jungfraujoch atlas (Delbouille and Roland, 1995). The two contrast profiles appear in absorption and are redshifted. It is very likely that the observed surge was located high above the mean chromosphere. Thus, it could be considered to be similar to a cloud illuminated from below. To derive the physical parameters of the surge, we applied the classical cloud model of radiative transfer (Beckers, 1964). Note that the cloud model is applicable for the absorption feature. In this case, the surge can be analyzed using the model. According to the model, the contrast of the observed intensity profile I_λ^{out} against its background intensity profile I_λ^{in} is described as

$$C_\lambda \equiv \frac{I_\lambda^{\mathrm{out}} - I_\lambda^{\mathrm{in}}}{I_\lambda^{\mathrm{in}}} = \left(\frac{S}{I_\lambda^{\mathrm{in}}} - 1\right)\left(1 - e^{-\tau_\lambda}\right), \tag{3}$$

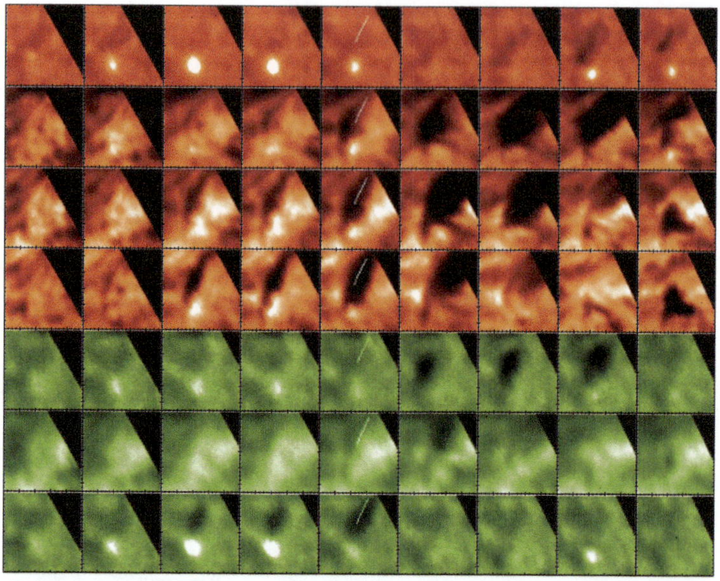

Figure 5 The time sequence of the Hα (top) and Ca II 8542 Å (bottom) lines in the red wing, center, and blue wing. The *x*-axis represents time in seconds from the first brightening of the EB. The field of view is $10 \times 10''$.

where S is the source function and τ_λ is the optical depth. We adopted τ_λ of the form

$$\tau_\lambda = \tau_{0,\lambda} \exp\left[-\left(\frac{\lambda - \lambda_0}{\Delta\lambda_D}\right)^2\right], \tag{4}$$

where λ_0 is the central wavelength of the absorption profile, $\Delta\lambda_D$ is the Doppler width of the line, and $\tau_{0,\lambda}$ is the optical depth at the line center. Note that λ_0 represents the LOS velocity v_0. The average over ten profiles taken at different times at the same location before the appearance of the surge was used for the background I_λ^{in}.

The cloud-model fitting was applied for the Hα and Ca II lines. Then λ_0, $\tau_{0,\lambda}$, S, and $\Delta\lambda_D$ were individually derived in two lines. The Doppler broadening $\Delta\lambda_D$ contains the thermal contribution and the nonthermal contribution. Then $\Delta\lambda_D$ is expressed as

$$\Delta\lambda_D = \frac{\lambda}{c}\sqrt{\frac{2kT}{M} + \xi^2}, \tag{5}$$

where, λ, T, M, ξ are the wavelength at rest, the kinetic temperature, the mass of the atom, and the speed of the nonthermal motion(traditionally called microturbulence), respectively. If the Hα and Ca II lines form in the same material, we can separately determine the values of T and ξ in the region from the two $\Delta\lambda_D$-values of the lines because the mass of the H I atom is different from that of the Ca II ion.

Figure 6 shows that the two profiles are fairly well fitted by the model. We find that the Hα contrast profile has a higher absorption contrast and broader width than the Ca II profile. Nevertheless, they have similar LOS velocity (about 6.8 $\mathrm{km\,s^{-1}}$), which strongly supports our expectation that the two contrast profiles originate from the same material.

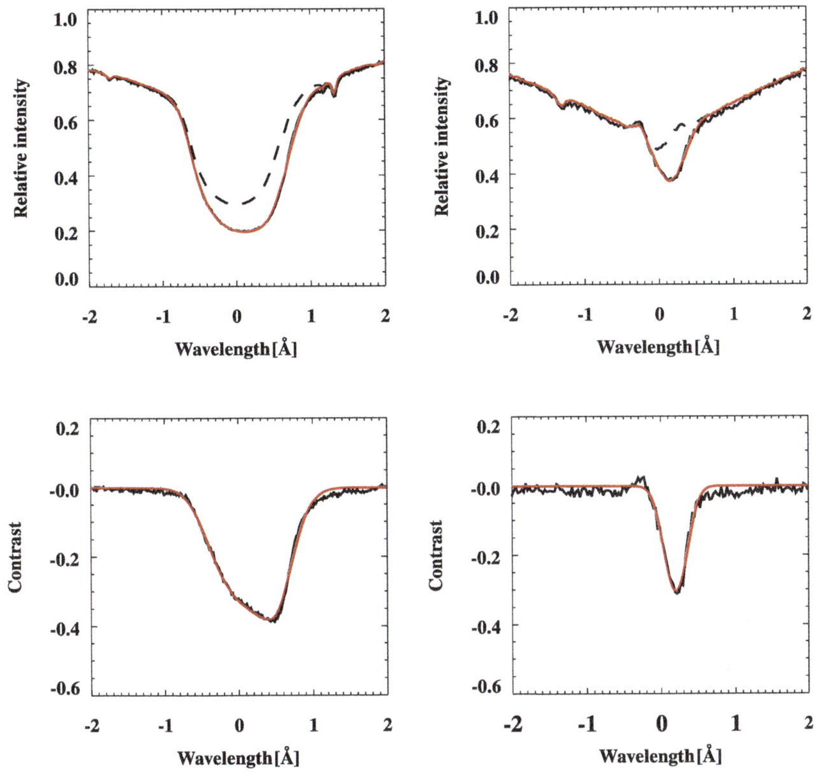

Figure 6 The intensity (top) and contrast (bottom) profiles of the surge in the Hα (left) and Ca II 8542 Å lines (right) at $t = 741$ s. The black dashed lines represent the intensity profiles of the background radiation ($I_\lambda^{\rm in}$), and the black solid lines are the data. The red solid lines are the cloud-model fitting results. The obtained LOS velocities for the Hα and the Ca II lines are $v_{\rm H\alpha} = 6.93$ km s^{-1}, and $v_{\rm Ca\,II} = 6.79$ km s^{-1}, respectively. The temperature and the nonthermal velocity are 24 700 K and 6.72 km s^{-1}, respectively.

Figure 7 shows the time-sequence maps of the LOS velocities T and ξ in each data set at the same FOV as shown in Figure 5. We manually masked the area of the surge by examining the images. After applying the fitting to the pixels inside the masked area, we discarded the points where the velocity difference between the Hα and Ca II lines was bigger than 5 km s^{-1}. As shown in the figure, the surge was obviously blueshifted in the early phase, then it changed to the redshift. Hα and Ca II lines show the same LOS velocities. The LOS velocity started to vary in the eastern part of the surge first, while the western part of the surge followed the variation. At the early phase of the surge, temperature T near the EB seems to be higher than the other part. Later, T at the middle of the surge increased to about 40 000 K. The value of ξ was higher than 15 km s^{-1} initially, but it decreased to about 5 km s^{-1} later.

We examined the temporal variation of physical parameters at 20 positions (about 3.6″) on the white curve marked in Figure 5. This curve follows the direction of the surge. At every position, we analyzed 50 profiles taken at different times during $t = 200$ s – 1200 s. Figure 8 shows the time-slice maps of the LOS velocities of the surge determined along the curve. We displayed values only at positions where the velocity difference between the Hα and Ca II lines was smaller than 5 km s^{-1}. The correlation between the chosen values of

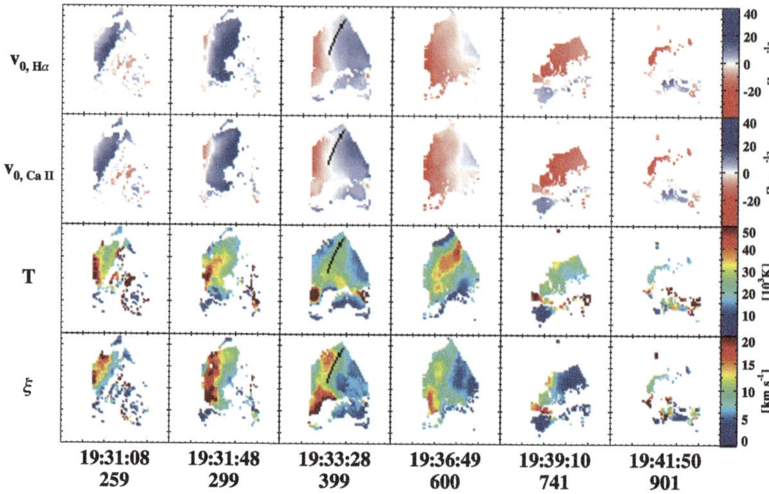

Figure 7 The time-sequence maps of the LOS velocities of Hα and Ca II 8542 Å lines, temperature and nonthermal velocity. The FOV is the same as in Figure 5.

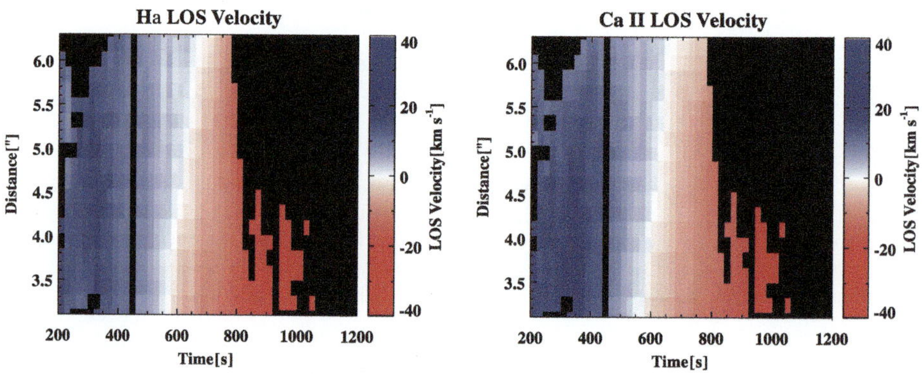

Figure 8 The time-slice maps of the LOS velocity of the surge calculated using the non-linear square fitting of the cloud model, in the Hα (left panel) and Ca II 8542 Å (right panel) lines. The positive value represents the upward motion.

Hα and Ca II is 0.99, and the standard deviation of $v_{\mathrm{H}\alpha} - v_{\mathrm{Ca\,II}}$ is 1.4 km s^{-1}. The velocity at $t = 200$ s was about 20 km s^{-1} upward. This blueshift simultaneously appeared at all positions, then they gradually changed and became redshifted. The estimated acceleration was almost constant (≈ -70 m s^{-2} downward) during the event, which is much smaller than the free-fall acceleration (-274 m s^{-2}) on the solar surface.

Figure 9 shows that the distributions of T and ξ vary over time. As seen in the figure, T ranged from 20 000 K to 40 000 K, and ξ from 5 km s^{-1} to 15 km s^{-1}. The mean and 1-σ values were found to be around $29\,000 \pm 4700$ K and 11.2 ± 1.9 km s^{-1}, respectively. Higher T values are found near the middle of the surge and higher ξ near its bottom. We found that T and ξ are moderately anti-correlated with a coefficient of -0.46 near the EB. Furthermore, we found a tendency for T to be as high as 40 000 K before $t = 800$ s, and

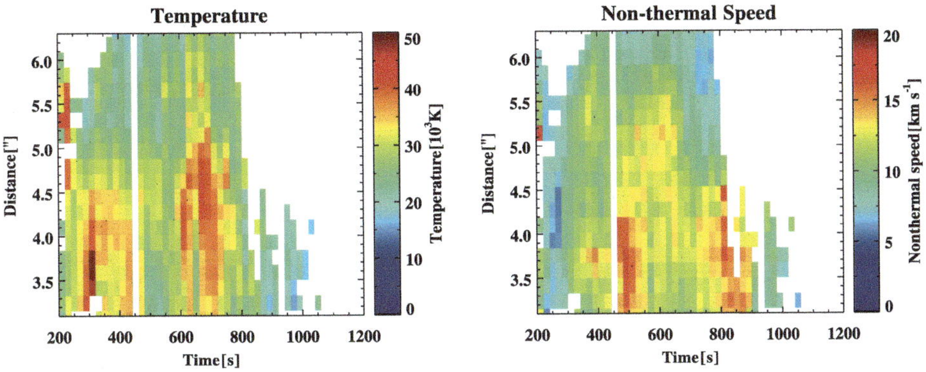

Figure 9 The time-slice maps of temperature (left panel) and nonthermal velocity (right panel) of the surge.

Figure 10 Temporal variation of (a) the LOS velocity of the surge at 3.1″distance from the EB, (b) the intensity at 6562.8 – 1.4 Å, (c) LOS velocity of emission wings of the EB, and (d) the transverse velocity under the EB. The LOS velocity of the surge is measured at 3.1″far from the EB. The transverse velocity is measured following the lane of the expanding granular cell. The red arrows in (b) represent the second and third EBs.

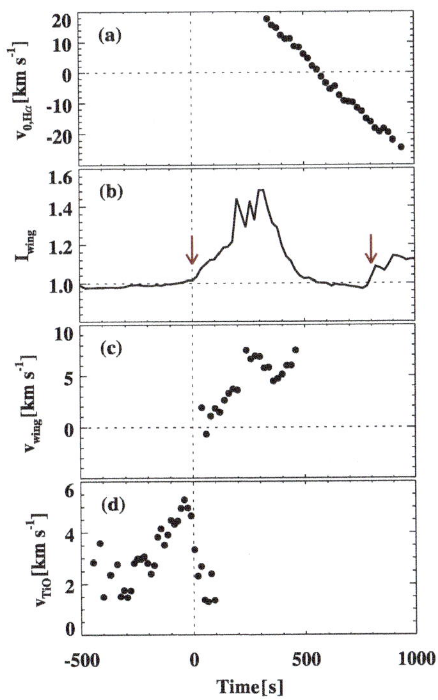

then T dropped to 20 000 K later. We also found a similar tendency that ξ decreased from 15 km s^{-1} to 5 km s^{-1}.

3.5. Summary of the Time Variation

The time variations of the velocities as well as the EB brightness are shown in Figure 10. It is evident that the speed of the photospheric transverse motion (v_{TiO}) steadily increased and then decreased in a short time. The time of the transition from acceleration to deceleration coincides with the beginning of the EB, suggesting that they were physically related. The low-chromosphere velocity measured from the emission wings (v_{wing}) increased during the

EB event, with an indication of a superposed oscillatory pattern at around $t = 400$ s. This is a little before the instant when the surge was first visible, moving at a high speed of about $20\ \mathrm{km\,s^{-1}}$. The LOS velocity of the surge ($v_{0,\mathrm{H}\alpha}$) monotonically decreased with an almost uniform deceleration of $\approx 70\ \mathrm{m\,s^{-2}}$.

4. Discussion

We have measured different kinds of velocity fields in relation to the EB as a function of time: the downflow and the transverse motion at the photospheric level, the upflow in the chromosphere, and the surge in the upper chromosphere or in the low corona. Moreover, we have obtained some results related to the heating process of the EB. We observed a brightening in the TiO images at the location of the EB. The intensity enhancement of the wings of the Ca II line indicates that the EB had a temperature excess of 200 K in the photosphere. The average temperature of the surge that originated from the EB was estimated to be about 29 000 K.

Our results help us to infer a comprehensive and detailed picture of various phenomena associated with the EB in the widely accepted frame of magnetic reconnection. In the photosphere, we observed events of brightening and downflow at the time of EB, and we also observed the upflow in the chromosphere. It is likely that these resulted from the magnetic reconnection that was responsible for the EB. If so, our observations suggest that the magnetic reconnection occurred between the upflow and the downflow regions – probably right above the photosphere, in agreement with previous studies (Georgoulis *et al.*, 2002; Pariat, Aulanier, and Schmieder, 2004; Watanabe *et al.*, 2011).

The observed magnetograms suggest that a magnetic bipole existed across the granular cell, even though the negative polarity was buried in the background positive pole (the resolution of the HMI is $0.91''$ (Schou *et al.*, 2012)), and expanded together with the cell (see Figure 2). Therefore, the transverse bi-directional motions of the lanes might represent an emerging bipole, and the ends of the expanding granule might be the footpoints of the bipole. It is very likely that the EB was located above the photospheric magnetic neutral line and was driven by the collision of the emerging negative flux with the pre-existing positive flux. The event looks similar to the emergence of an Ω-shaped flux loop observed by Lim *et al.* (2011).

Our observations indicate the driving force of the reconnection: it could be a small-scale flux emergence that manifested itself in an expanding granule. One end point of the granule – supposedly one footpoint of the emerging field – moved at a fast speed and eventually reached a point that was very close to the EB site, suggesting that the reconnection process responsible for this EB may have been driven by this flux emergence. The small distance between the end point of the granule and the EB site also suggests that the reconnection occurred at a low altitude. The observed motion toward the site of the EB was as fast as $6\ \mathrm{km\,s^{-1}}$ and hence is quite different from the previously reported continuous, wide horizontal flows with a speed of $0.1-0.3\ \mathrm{km\,s^{-1}}$ that were suspected to be inflows to the EB site (Georgoulis *et al.*, 2002). Rapidly converging flows of $4-6\ \mathrm{km\,s^{-1}}$ have been observed in recent high spatial resolution observations (Jess *et al.*, 2010; Lim *et al.*, 2011; Chae *et al.*, 2010). It is surprising that the speeds are as fast as the sound speed of the photospheric level. The transverse motion changed from acceleration to deceleration when the one end point of the granule reached the EB site. This rapid change may result from the high magnetic pressure of the flux pile-up site, which resists against further convergence of the magnetic elements and frozen-in plasma. At any rate, the fact that the EB started after the end point of the granule reached the site strongly suggests that the transverse motion may have driven the process of magnetic reconnection, even though flux emergence is not seen in our data.

We also found a systematic difference in the profile shape of the EB between the Hα line and Ca II line. Unlike the Hα line, the spectral characteristics of the broad wing emission are not clear in Ca II. The Ca II 8542 Å line is strongly affected by the collisional effect because the line shares the upper level with the collisionally excited line, and hence reflects the temperature of the line formation region. In contrast, the Hα line is dominated by radiative processes such as radiative excitation/de-excitation and photoionization/radiative recombination, and hence is insensitive to the local kinetic temperature in the chromosphere (Cauzzi *et al.*, 2009; Chae *et al.*, 2012). Therefore, the above spectral characteristics strongly suggest that the broad emission wings of the Hα line in the EB may have a nonthermal origin, possibly an energetic-particle beam (Ding, Henoux, and Fang, 1998).

Even though we were unable to resolve the EB in our observations, recent high-resolution observations such as reported by Watanabe *et al.* (2011) indicate that the EB has a shape of an anemone jet. The speeds of the upflow in the EB we observed and those of previous measurements (Kitai, 1983; Kurokawa *et al.*, 1982) are consistent with this new picture. If the material moves upward for 500 s with the speed of $1-7\ \mathrm{km\,s^{-1}}$ as shown in Figure 10, it can reach a height of about 2500 m, which is higher than the typical EB size of $1.8''$ (1300 km).

Our results may contribute to a better understanding of the dynamics and energetics of the surge that originates from the EB. Note that the surge had a speed of $20\ \mathrm{km\,s^{-1}}$ from the first instant when it emerged as a clear dark feature. The speed of the surge at the earlier times may have been even higher than this value because it was already decelerating at the instant of the observation. Choosing $\Delta t = 200$ s for the duration of the initial acceleration, and $v = 25\ \mathrm{km\,s^{-1}}$ for the peak speed, we obtained $-125\ \mathrm{m\,s^{-2}}$ as an estimate of the average deceleration. The temperature and speed of the surge we obtained appear to be consistent with the observation of Tziotziou, Tsiropoula, and Sütterlin (2005), which yielded a temperature of $10^4 - 10^5$ K, and the simulation of Yokoyama and Shibata (1996), which yielded a speed of a few tens of $\mathrm{km\,s^{-1}}$.

Our finding that the surge has temperatures ranging from 20 000 to 40 000 K raises an important question to be addressed; how can the surge at such high temperatures be seen in the Hα and Ca II absorption? First, why in absorption, not in emission? This is because the scattering process dominates in the source functions of the lines in the low-density, high-lying features. Temperature is not that important in this system governed by scattering processes. Second, how can neutral hydrogen exists at such high temperatures? This is a serious problem. It is well-known that in LTE, most hydrogen atoms are ionized at temperatures above 20 000 K. This behavior appears to hold even in non-LTE because we find that temperatures inside filaments/prominences are measured to be lower than 20 000 K (Park *et al.*, 2013; Song, Henoux, and Fang, 2013). One possible explanation for this discrepancy is that the surge material may not be in ionization equilibrium. This kind of deviation from equilibrium arises when the ionization/recombination timescales ($10^3 - 10^5$ s in the chromosphere according to Carlsson and Stein (2002)) are longer than the dynamic timescales, such as in the surge in which the dynamic timescale is a few hundred seconds, as we observed.

In conclusion, our observations support the picture that an EB and its associated surge is a response of the atmosphere to the process of magnetic reconnection that is driven by small-scale flux emergence and occurs in the low chromosphere.

Acknowledgements We greatly appreciate the referee's constructive comments. We are grateful to the BBSO team for their technical support for the observations. This work was supported by the National Research Foundation of Korea (2011-0028102). E.-K. Lim is supported by the "Study of near-Earth effects by CME/HSS" project and basic research funding from KASI. Yeon-Han Kim and Eun-Kyung Lim were supported by the Construction of Korean Space Weather Center, the project of KASI, and the KASI basic research fund.

References

Altschuler, M.D., Lilliequist, C.G., Nakagawa, Y.: 1968, A possible acceleration mechanism for a solar surge. *Solar Phys.* **5**, 366 – 376. doi:10.1007/BF00147150.

Beckers, J.M.: 1964, A study of the fine structures in the solar chromosphere. Ph.D. thesis, Sacramento Peak Observatory, Air Force Cambridge Research Laboratories, USA.

Canfield, R.: 1996, Hα surges and X-ray jets in AR 7260. *Astrophys. J.* **464**, 1016 – 1029. doi:10.1086/7389.

Cao, W., Gorceix, N., Coulter, R., Ahn, K., Rimmele, T.R., Goode, P.R.: 2010, Scientific instrumentation for the 1.6 m New Solar Telescope in Big Bear. *Astron. Nachr.* **331**. doi:10.1002/asna.201011390.

Carlsson, M., Stein, R.F.: 2002, Dynamic hydrogen ionization. *Astrophys. J.* **572**, 626 – 635. doi:10.1086/340293.

Cauzzi, G., Reardon, K., Rutten, R.J., Tritschler, A., Uitenbroek, H.: 2009, The solar chromosphere at high resolution with IBIS. IV. Dual-line evidence of heating in chromospheric network. *Astron. Astrophys.* **503**, 577 – 587. doi:10.1051/0004-6361/200811595.

Chae, J., Sakurai, T.: 2008, A test of three optical flow techniques – LCT, DAVE, and NAVE. *Astrophys. J.* **689**, 593 – 612. doi:10.1086/592761.

Chae, J., Goode, P.R., Ahn, K., Yurchysyn, V., Abramenko, V., Andic, A., Cao, W., Park, Y.D.: 2010, New Solar Telescope observations of magnetic reconnection occurring in the chromosphere of the quiet Sun. *Astrophys. J. Lett.* **713**, L6 – L10. doi:10.1088/2041-8205/713/1/L6.

Chae, J., Park, H.-M., Ahn, K., Yang, H., Park, Y.-D., Nah, J., Jang, B.H., Cho, K.-S., Cao, W., Goode, P.R.: 2012, Fast Imaging Solar Spectrograph of the 1.6 meter New Solar Telescope at Big Bear Solar Observatory. *Solar Phys.* doi:10.1007/s11207-012-0147-x.

Dara, H.C., Alissandrakis, C.E., Zachariadis, T.G., Georgakilas, A.A.: 1997, Magnetic and velocity field in association with Ellerman bombs. *Astron. Astrophys.* **322**, 653 – 658.

Delbouille, L., Roland, G.: 1995, High-resolution solar and atmospheric spectroscopy from the Jungfraujoch high-altitude station. *Opt. Eng.* **34**, 2736 – 2739. doi:10.1117/12.205676.

Ding, M.D., Henoux, J.-C., Fang, C.: 1998, Line profiles in moustaches produced by an impacting energetic particle beam. *Astron. Astrophys.* **332**, 761 – 766.

Ellerman, F.: 1917, Solar hydrogen "bombs". *Astrophys. J.* **46**, 298 – 300. doi:10.1086/142366.

Fang, C., Tang, Y.H., Xu, Z., Ding, M.D., Chen, P.F.: 2006, Spectral analysis of Ellerman bombs. *Astrophys. J.* **643**, 1325 – 1336. doi:10.1086/501342.

Georgoulis, M.K., Rust, D.M., Bernasconi, P.N., Schmieder, B.: 2002, Statistics, morphology, and energetics of Ellerman bombs. *Astrophys. J.* **575**, 506 – 528. doi:10.1086/341195.

Isobe, H., Tripathi, D., Archontis, V.: 2007, Ellerman bombs and jets associated with resistive flux emergence. *Astrophys. J. Lett.* **657**, L53 – L56. doi:10.1086/512969.

Jess, D.B., Mathioudakis, M., Browning, P.K., Crockett, P.J., Keenan, F.P.: 2010, Microflare activity driven by forced magnetic reconnection. *Astrophys. J. Lett.* **712**, L111 – L115. doi:10.1088/2041-8205/712/1/L111.

Kitai, R.: 1983, On the mass motions and the atmospheric states of moustaches. *Solar Phys.* **87**, 135 – 154. doi:10.1007/BF00151165.

Kurokawa, H., Kawaguchi, I., Funakoshi, Y., Nakai, Y.: 1982, Morphological and evolutional features of Ellerman bombs. *Solar Phys.* **79**, 77 – 84. doi:10.1007/BF00146974.

Lim, E.-K., Yurchyshyn, V., Abramenko, V., Ahn, K., Cao, W., Goode, P.: 2011, Photospheric signatures of granular-scale flux emergence and cancellation at the penumbral boundary. *Astrophys. J.* **740**, 82. doi:10.1088/0004-637X/740/2/82.

Matsumoto, T., Kitai, R., Shibata, K., Otsuji, K., Naruse, T., Shiota, D., Takasaki, H.: 2008, Height dependence of gas flows in an Ellerman bomb. *Publ. Astron. Soc. Japan* **60**, 95 – 102.

Pariat, E., Aulanier, G., Schmieder, B.: 2004, Resistive emergence of undulatory flux tubes. *Astrophys. J.* **614**, 1099 – 1112. doi:10.1086/423891.

Pariat, E., Schmieder, B., Berlicki, A., Deng, Y., Mein, N., Ariste, A.L., Wang, S.: 2007, Spectrophotometric analysis of Ellerman bombs in the Ca II, Hα, and UV range. *Astron. Astrophys.* **289**, 279 – 289. doi:10.1051/0004-6361:20067011.

Park, H., Chae, J., Song, D., Maurya, R., Yang, H., Park, Y., Jang, B., Nah, J., Cho, K., Kim, Y., Ahn, K., Cao, W., Goode, P.: 2013, Determination of temperature of solar prominences/filaments using Fast Imaging Solar Spectrograph with the 1.6 m New Solar Telescope at the Big Bear Solar Observatory – I. Limb. *Solar Phys.* doi:10.1007/s11207-013-0271-2.

Payne, T.E.W.: 1993, A multiwavelength study of solar Ellerman bombs. Ph.D. thesis, New Mexico State University.

Roy, J.R.: 1973, The magnetic properties of solar surges. *Solar Phys.* **28**, 95 – 114. doi:10.1007/BF00152915.

Rust, D.: 1968, Chromospheric explosions and satellite sunspots In: Kiepenheuer, K.O. (ed.) *Structure and Development of Solar Active Regions*, *IAU Symp.* **35**, 77 – 84.

Schou, J., Scherrer, P.H., Bush, R.I., Wachter, R., Couvidat, S., Rabello-Soares, M.C., Bogart, R.S., Hoeksema, J.T., Liu, Y., Duvall, T.L., Akin, D.J., Allard, B.A., Miles, J.W., Rairden, R., Shine, R.A., Tarbell, T.D., Title, A.M., Wolfson, C.J., Elmore, D.F., Norton, A.A., Tomczyk, S.: 2012, Design and ground calibration of the Helioseismic and Magnetic Imager (HMI) instrument on the Solar Dynamics Observatory (SDO). *Solar Phys.* **275**, 229 – 259. doi:10.1007/s11207-011-9842-2.

Song, D., Henoux, J.-C., Fang, C.: 2013, Determination of temperature of solar prominences/filaments using Fast Imaging Solar Spectrograph with the 1.6 m New Solar Telescope at the Big Bear Solar Observatory – II. Filament. *Solar Phys.* submitted.

Tziotziou, K., Tsiropoula, G., Sütterlin, P.: 2005, DOT tomography of the solar atmosphere. V. Analysis of a surge from AR10486. *Astron. Astrophys.* **444**, 265 – 274. doi:10.1051/0004-6361:20053604.

Watanabe, H., Vissers, G., Kitai, R., Voort, L.R.V.D., Rutten, R.J.: 2011, Ellerman bombs at high resolution: I. Morphological evidence for photospheric reconnection. *Astrophys. J.* **736**, 71. doi:10.1088/0004-637X/736/1/71.

Wöger, F., von der Lühe, O., Reardon, K.: 2008, Speckle interferometry with adaptive optics corrected solar data. *Astron. Astrophys.* **488**, 375 – 381. doi:10.1051/0004-6361:200809894.

Yokoyama, T., Shibata, K.: 1996, Numerical simulation of solar coronal X-ray jets based on the magnetic reconnection model. *Publ. Astron. Soc. Japan* **48**, 353 – 376.

DOI 10.1007/978-3-319-12123-9_4
Reprinted from *Solar Physics* Journal, DOI 10.1007/s11207-012-0113-7

Connection Between Chromospheric Events and Photospheric Dynamics

A. Anđić · J. Chae · H. Park · H. Yang · K. Ahn · W. Cao · Y.D. Park

Received: 29 June 2011 / Accepted: 20 August 2012 / Published online: 16 October 2012
© Springer Science+Business Media B.V. 2012

Abstract We analyzed chromospheric events and their connection to oscillation phenomena and photospheric dynamics. The observations were done with the New Solar Telescope of Big Bear Solar Observatory using a broad-band imager at the wavelength of a TiO band and FISS spectrograph scanning Ca II and Hα spectral lines. The event in Ca II showed strong plasma flows and propagating waves in the chromosphere. The movement of the footpoints of flux tubes in the photosphere indicated flux tube entanglement and magnetic reconnection as a possible cause of the observed brightening and waves propagating in the chromosphere. An upward propagating train of waves was observed at the site of the downflow event in Hα. There was no clear relationship between photospheric waves and the Ca II and Hα events. Our observations indicate that chromospheric waves that were previously thought to originate from the photosphere may be generated by some events in the chromosphere as well.

Keywords Chromosphere · Photosphere

1. Introduction

Recent research on the solar atmosphere has demonstrated complexities of the chromosphere, showing us that there is still a plethora of unexplained and previously unresolved

Initial Results from FISS
Guest Editor: Jongchul Chae

A. Anđić (✉)
Department of Astronomy, New Mexico State University, P.O. Box 3001, MSC 4500, Las Cruces, NM 88003, USA
e-mail: andic@astronomy.nmsu.edu

A. Anđić · K. Ahn · W. Cao
Big Bear Solar Observatory, 40386 North Shore Lane, Big Bear City, CA 92314, USA

J. Chae · H. Park · H. Yang · Y.D. Park
Astronomy Program, Department of Physics and Astronomy, Seoul National University, Seoul 151-741, Korea

dynamics there. Starting with Ellerman (1917) who noted a sudden bright event in the Hα line (Ellerman's bomb), the research so far has connected Ellerman bombs and K_{2V} brightenings to magnetic elements (Sivaraman and Livingston, 1982; Nindos and Zirin, 1998; Sivaraman *et al.*, 2000).

The complexity of the Ellerman bombs slowly emerged in the course of the research. Oscillatory signatures were observed in magnetic flux tubes by Volkmer, Kneer, and Bendlin (1995). From a theoretical standpoint Kalkofen (1997) interpreted intensity waves observed in the Ca II lines in terms of transverse and longitudinal magnetoacoustic waves propagating upward inside magnetic flux tubes. Successful models of Ca II grains, generated by acoustic shocks, were made by Carlsson and Stein (1992, 1997); they showed that bright Ca grains are produced primarily by waves from the photosphere whose frequency is slightly above the acoustic cut-off frequency. Connection between waves and brightening in the Ca bright points is also established and confirmed by observations; even later observations did not find any disagreement. Kariyappa, Narayanan, and Dame (2005) found that the chromospheric brightening observed in the Ca II H line is related to oscillations with a constant period (≈3 min), independent of their peak brightness.

Kalkofen (1999) proposed transverse and longitudinal waves in the magnetic flux tubes as an explanation for the observed dynamics, while Noble, Musielak, and Ulmschneider (2003) indicated the existence of torsional magnetic tube waves in flux tubes. The latter study was followed up by Musielak and Ulmschneider (2003) who considered the generation of longitudinal tube waves by the non-linear mode coupling with transverse waves and reproduced the observed 3-min oscillations. Khomenko, Collados, and Felipe (2008) found that horizontal motions in a deep portion of the flux tube generate a slow magnetosonic mode and a surface mode. These modes are efficiently transformed into a slow acoustic mode in the atmosphere where the Afvén speed is lower than the sound speed. Such a transformed mode might propagate along the field lines and effectively deposit the energy of the oscillatory driver into the chromosphere. These waves were observed two years later (Andić *et al.*, 2010). However, a complex role played by the magnetic field in the lower atmosphere still gives a plenty of research topics. De Moortel and Galsgaard (2006) investigated the magnetic reconnection driven by rotational footpoint motions. An observational case was reported by Jess *et al.* (2007) who found rotational movement of oscillation sites (which were identified with the footpoints of flux tubes) during microflaring events.

The subsequent research has uncovered complexity in the chromospheric waves. Bodo *et al.* (2001) found that the temperature difference in a plasma, either in the horizontal or vertical direction, is enough to change the direction of the upward propagating waves, making them propagate along curved paths. Rosenthal *et al.* (2002) found that waves in the magnetic field that is significantly inclined from the vertical direction tend to get reflected from the surface of varying altitude. Jain and Haber (2002) found that acoustic waves are suppressed by the magnetic field, and the amount of the suppression increases with the field strength. Moreover, Bogdan *et al.* (2003) demonstrated that wave propagation in the presence of the magnetic field is complex. Some waves come directly from the source while the other waves come from the magnetic canopy due to mode conversion. The propagation directions of individual wave trains are not co-aligned with each other nor with the observer's line of sight. Also the characteristics of the observed waves depend on the ratio of thermal to magnetic pressure of their environment. This trend seems to continue in recent research. Shelyag *et al.* (2009) indicated that the magnetic field perturbs and scatters acoustic waves and absorbs the acoustic power of a wave packet. Their model showed that a highly curved strong magnetic field also partially transforms the wave packet into the slow magnetosonic mode that propagates downwards.

All these research results so far indicate that there might be a possibility that not all of the Ca II grains are generated by acoustic shocks, especially the ones located near complex magnetic field topography. Moreover, there is indication that photospheric and chromospheric waves may be independently excited (Leibacher, Gouttebroze, and Stein, 1982). Also, recently Reardon *et al.* (2008) found that chromospheric turbulence is generated by the acoustic shocks that reach the appropriate height due to steep vertical density gradients and that non-linear shock processes produce the cascade of energy to higher frequencies.

In this work we present the observations of a chromospheric bright point and its connection to the photospheric dynamics. In Section 2 we present the data and explain analysis procedures. In Section 3 we present results; first introducing the events themselves and observed photospheric dynamics, then we analyze waves detected in both chromospheric events and in the photosphere. In Section 4 we discuss results and connect them to the previous work. In Section 5 we summarize the conclusions.

2. Data and Analysis

The observations were performed with the New Solar Telescope (NST) at Big Bear Solar Observatory (Goode *et al.*, 2010) on 12 July 2009, using two instruments. Seeing was average and the low order adaptive optics system was used.

Imaging data of a small pore at solar coordinates E450″ N248″ were obtained using the optical setup at the Nasmyth focus with the following components: a broad-band filter centered at an absorption band of TiO molecules (705.68 nm) with a 1 nm band pass. This filter has also a 1 nm leakage around 613 nm with 25 % of the transmission of the main transmission band. However, this does not jeopardize the interpretation of our observations. The transmission of the filter is broad and averages over the line and continuum contributions, making the signal only weakly dependent on the properties of individual spectral lines. This is desirable in our case since the TiO line is very weak in the quiet Sun. Hence, we can assume that the height of formation of the image data is very close to that of the continuum. The used detector was a PCO.2000 camera (Cao *et al.*, 2010). The data set covers the time sequence of half an hour with 120 bursts of exposure, each of which was composed of 100 frames. The exposure of individual frames was 1 ms, and the cadence between bursts was 15 s. The field of view (FOV) was 70″ × 70″. The achieved resolution by Dawes' limit is 0.09″.

Simultaneously we performed spectroscopic observations with the Fast Imaging Solar Spectrograph (FISS) (Chae *et al.*, 2010). FISS is a field-scanning slit spectrograph with a typical spectral resolution of 1.4×10^5 and a unique capability for simultaneous observations at Hα and Ca II 854 nm in the fast scanning mode. For the observations reported here we scanned the field of 16″ × 40″, achieving a temporal cadence of 18 s. This FOV was contained within the FOV of the TiO imaging data. FISS was scanning both lines simultaneously with the spatial sampling of 0.16″. The angular resolution was determined by the seeing condition and is estimated to be about 1″. The images from the two instruments were co-aligned using a small pore in the FOV, since the resolution of the FISS instrument does not reveal the same detail in the photosphere as the TiO-band images. The pore itself was 3″ away from the event.

The TiO data set was calibrated (corrected for dark frames and flat-fielded) and then speckle reconstructed based on the speckle masking method (von der Lühe, 1993). For this purpose we used the Kiepenheuer Institut Speckle Interferometry Package (Wöger, von der Lühe, and Reardon, 2008). The cadence of the reconstructed TiO data provided us with a

Nyquist frequency of 67 mHz. The reconstructed images were aligned by a cross-correlation technique using a Fourier-transform routine and squared mean absolute deviations to provide sub-pixel alignment accuracy. We did not perform sub-pixel image shifting to avoid the interpolation errors that are inevitable with the use of this technique.

FISS data sets were also corrected for dark frames and flat-fielded following the procedures described by Chae *et al.* (2010). Velocity maps were constructed from the FISS data set using two different methods; the center-of-gravity method (Janssen, 2003) and the bisector method (Andjic, 2006; Andić, 2007a, 2007b). For both methods the reference (zero) point of the velocity was the average value of the analyzed points in the spectral line over the whole data set.

A wavelet analysis of intensity and velocity data was performed using the Morlet wavelet with the code based on Torrence and Compo (1998). An automated procedure was used here to carry out the wavelet analysis, as has been carried out previously and presented in detail by Bloomfield *et al.* (2006), with the restrictions described in detail in Andić *et al.* (2010).

As Lites and Chipman (1979) established, in a pure acoustic wave the upward velocity is in phase with enhancements in pressure, temperature, and density. Therefore, the phase relationship between the velocity and intensity signals in a spectral line can give insight into the nature of the observed waves and propagation characteristics. We analyzed the phase coherence using the wavelet approach (Bloomfield *et al.*, 2004). The difference in power contained in a time series computed with the traditional Fourier analysis and the wavelet analysis is small. The Fourier analysis, without timing information, has similar confidence levels when compared to the wavelet analysis averaged over time. Since the phase relation between the wave packets can change in time, the Fourier analysis loses information on these changes due to lack of timing information. The major cause of this loss is the fact that the phase can take a negative value as well, while the power is always positive. This is one reason why the wavelet analysis was adopted in our analysis. Since for pure noise this procedure yields positive coherence, only results with a coherence above 0.6 are regarded as significant.

The phase difference and phase coherence were calculated using the intensity and velocity data of two spectral line cores obtained with FISS and using intensity signal from the TiO data set. The velocity data were obtained from the spectral line bisector shift (Andjic, 2006; Andić, 2007a, 2007b). This processing allowed us to calculate phase differences in velocity–velocity (V-V), velocity–intensity (V-I), and intensity–intensity (I-I) data. The phase differences were calculated using the same spatial location (*i.e.* the same pixel in the field of view). The V-V and I-I phase differences were derived from two different spectral lines, while the V-I phase differences were calculated from the same spectral line. Therefore, the V-V and I-I phase differences reflect the difference in formation height of the corresponding line cores (Hα and Ca II).

Since the TiO data have only intensity images, in order to estimate the oscillatory phase of waves a different method had to be used. We combined the Fourier and Hilbert transforms on the single data set (White and Cha, 1973). The resulting signal is a complex function; its real part is the original signal and its imaginary part is a quadrature of the original signal. Since a real function and its quadrature are Hilbert transform pairs, the Hilbert transform converts one into another. The resulting transform describes the amplitude and phase of a variable in the complex plane. The signal is transformed into the Fourier space then transformed back using the Hilbert transform (Stebbins and Goode, 1987).

3. Results

We observed a brightening in the Ca II 854 nm line. In the vicinity of the brightening area, on a granular scale distance, we observed strong downflow in the Hα 656.28 nm line. The Ca II brightening and the Hα downflow areas were separated by about 3″. The events appeared simultaneously in Ca II 854 nm (Figure 1B) and Hα (Figure 1C), but at different locations. The events in both lines lasted for about 3 min. Prior to the Ca II brightening, we observed motions of resolved bright points (BPs) in the lower photosphere directly below the brightening area. These motions appeared at 8 min from the beginning of our time series and lasted until 12 min.

3.1. Line Profiles and Velocities

The velocity maps were obtained for both lines. For the Hα line, velocity maps were made from the wavelengths at the center of the line (656.28 nm), blue wing (656.21 nm), red wing (656.21 nm), blue continuum (655.9 nm), and red continuum (656.69 nm). For the Ca II line they were made from the wavelengths at the line center (854.21 nm), blue wing (854.17 nm), red wing (854.25 nm), blue continuum (853.65 nm), and red continuum (854.73 nm). The velocity maps revealed that the event in Hα was connected with strong downflow of plasma, while the brightening in Ca II with strong plasma upflow.

Figure 1C represents a strong downflow event observed in the velocity maps of Hα at spatial coordinates (5.6″, 4.1″). The duration of the event was determined by intensity changes in the area. The velocity signal at this location showed constant downflow, with its maximum velocity appearing ≈0.5 minute after the intensity showed the minimum value. The Hα spectral line profiles of this event show a red shift indicating velocities of about 5 $\mathrm{km\,s^{-1}}$ (Figure 2). The intensity at the red wing is 16 % higher than the one at the blue wing, indicating downflow. At the location of the Ca II event, downflow was observed in the Hα as well.

The simultaneous event in the Ca II line, separated from the Hα event by a spatial distance comparable to the granule size, showed indications of upflow (Figure 1B). The Ca II line profiles show a slight blue shift that corresponds to velocities of 2 $\mathrm{km\,s^{-1}}$ (Figure 3). The Ca II event is located at (1.4″, 2″) of the FOV. This event showed increases in velocity and intensity over the duration of the observation. The velocities measured at the wings were 30 % slower than the velocities observed at the line core. The Ca II velocity maps made from the line core and the blue wing both show the downflow at the location of the Hα event, which indicated that the same event was visible in both lines.

3.2. Photospheric Dynamics Under the Hα Event

Previous researches have indicated connections between Ca II brightening and magnetic elements (Sivaraman and Livingston, 1982; Nindos and Zirin, 1998; Sivaraman *et al.*, 2000). Our present analysis, on the contrary, suggests a possible connection between this chromospheric event and the dynamics of BPs in the photosphere (Figure 1A). In the following we describe the observed motions of small structures in the photosphere that were located below the corresponding chromospheric events.

In order to study the dynamics of BPs in the photosphere below the Hα event, we have selected a subregion of 3″ × 3″ in the TiO images. The Hα spectral data indicate downflow in this area, with its speed increasing throughout the observing run. This behavior in the Hα data suggests that the Hα downflow event might be associated with pronounced intergranular lanes in the TiO images at the same location, in view of the fact that intergranular lanes

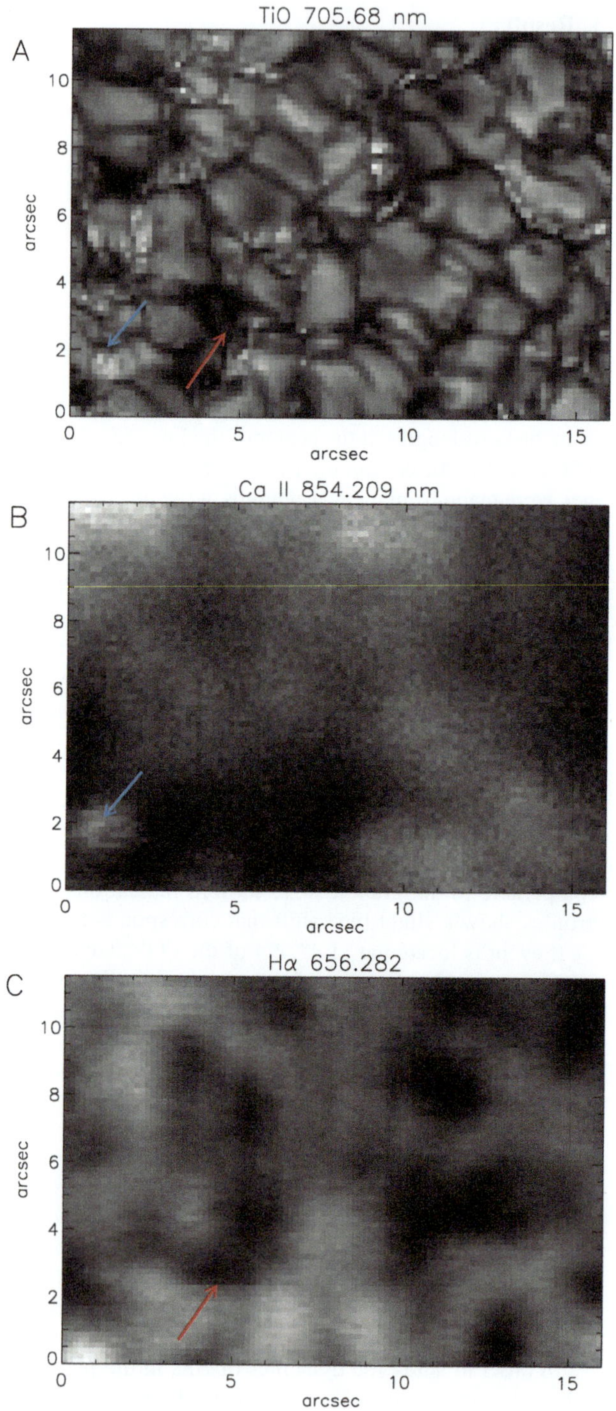

Figure 1 Observed events and corresponding photospheric configuration. Panel A is an image taken with a broad-band filter centered at the TiO band (705.68 nm). Arrows in panel A indicate small structures related to the event under consideration. The red and the blue arrows point to the locations of the Hα event and of the Ca II event, respectively. Panel B represents a core intensity image of the Ca II line 9 min from the beginning of the time series with the blue arrow marking the event, in this case brightening. Panel C is an Hα line core intensity image at the same moment as the peak intensity of the Ca II event. The red arrow points to the darkening event seen in this frame. Panels B and C were obtained by the FISS instrument.

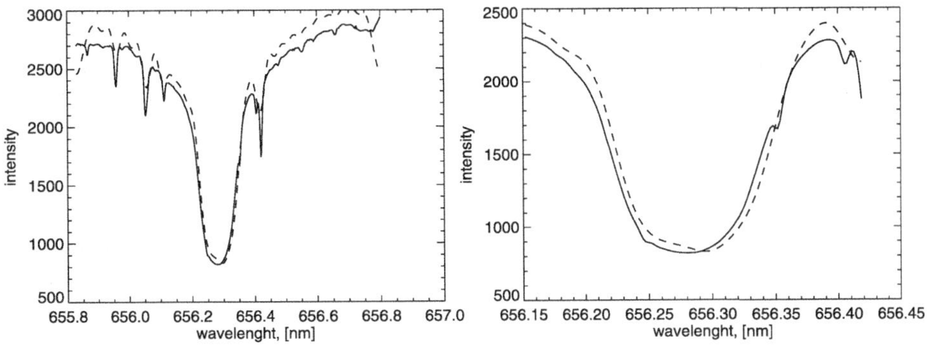

Figure 2 Hα line profiles of the Hα event. The left panel shows the full profile, while the right panel shows its enlarged line core. Solid lines represent a reference profile obtained by averaging the profiles over the whole field of view (FOV). Dashed lines are example profiles from the event. Noise in the profiles was filtered out with a low pass Fourier filter. A red shift corresponding to 6 km s^{-1} is seen.

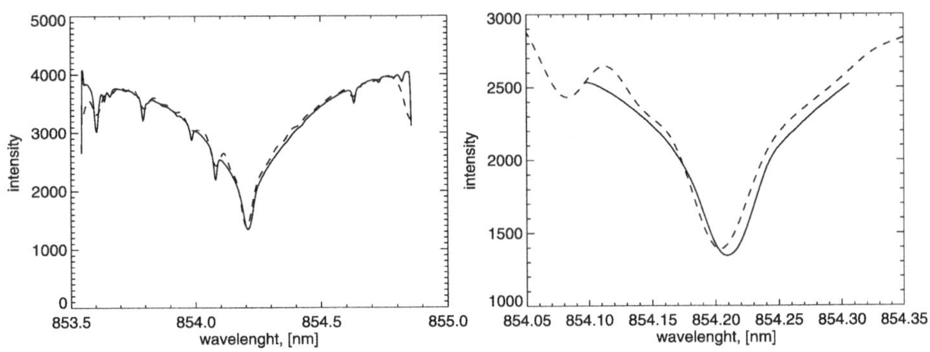

Figure 3 Ca II line profile of the Ca II event. The left panel shows the full profile, while the right panel shows its enlarged line core. Solid lines represent a reference profile obtained by averaging the profiles over the whole FOV. Dashed lines are example profiles from the event. Noise in the profiles was filtered out using a low pass Fourier filter. A blue shift corresponding to 2 km s^{-1} is seen.

are the places of persistent photospheric downflow. In fact, such an intergranular lane was observed as shown in Figure 4. A very broad lane started to form at 2 min after the beginning of the time series, and lasted till 5.5 min, when it contracted to a size typical of intergranular lanes. This continued presence of the intergranular lane during the Hα downflow event indicates a potential connection between chromospheric and photospheric dynamics.

In this intergranular lane we identified five BPs, and tracked them using the NAVE method (Chae and Sakurai, 2008) across the FOV of the subregion. As shown in Figure 5, the path of each BP neither crossed nor got closer to others' paths during the observing duration. This clear separation of the paths indicates that the motions of these BPs did not involve any flux tube entanglement at the photospheric level.

3.3. Photospheric Dynamics Under the Ca II Event

The photosphere below the Ca II brightening event showed different characteristics (Figure 6). In this location 21 BPs were identified and the majority of them were visible throughout the whole time series. But six of them were short-lived and ended their existence close

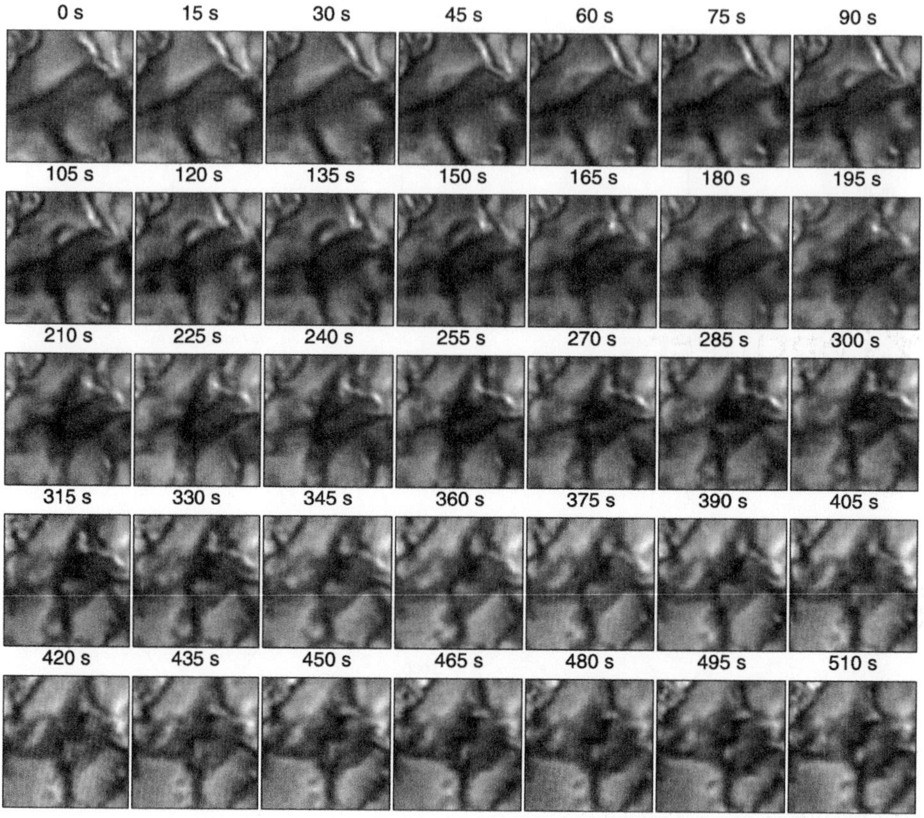

Figure 4 Time series of TiO filter images of the photosphere directly below the Hα event. Individual frame size is 3″ × 3″ and the time of observation measured from the beginning of the time series is shown above each frame.

Figure 5 Average TiO-band image of the photosphere below the Hα event, taken by integrating over the first 10 min of the observing run. Colored lines in the frame represent the paths of five detected BPs during the observing run moving toward left.

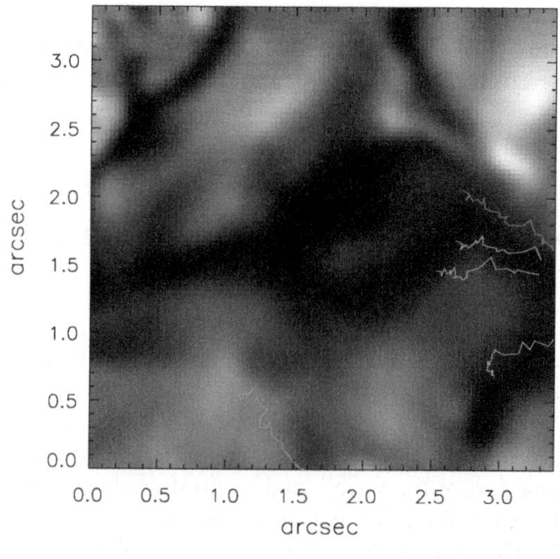

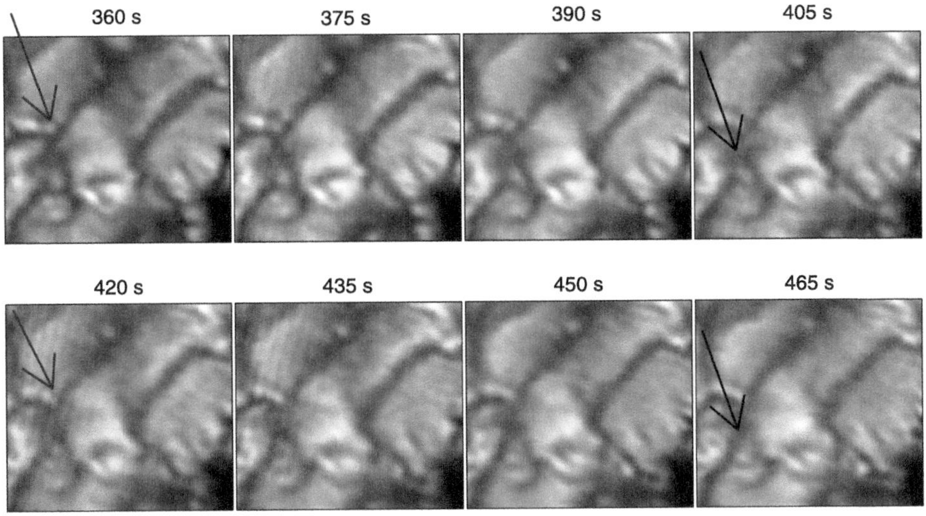

Figure 6 Time series of TiO filter images of the photosphere directly below the Ca II event. Individual frame size is $3'' \times 3''$. The arrows point to two pairs of BPs of our interest. The red arrows indicate one pair of BPs that was visible until 435 s, and the blue arrows indicate another pair of BPs that survived until 480 s.

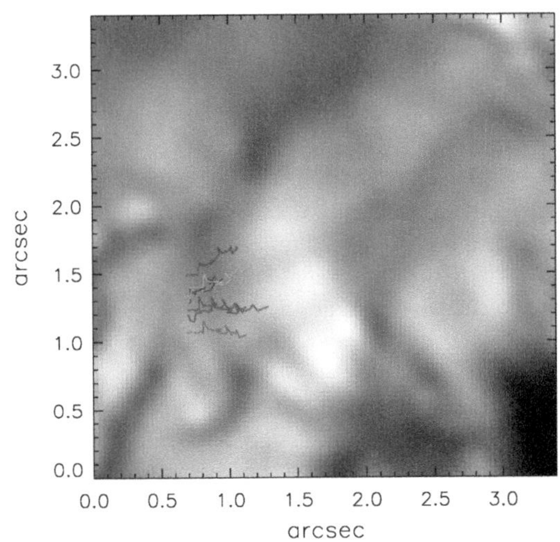

Figure 7 Average TiO-band image of the photosphere below the Ca II event, taken by integrating over the first 10 min of the observational run. Colored curves in the frame represent the trajectories of six BPs moving towards left.

in time to the moment of maximum intensity and velocity amplitude of the Ca II upflow event. The disappearance occurred in pairs. One pair of BPs (as marked by the red arrows in Figure 6) ended their life at 6.75 min of the time series, and the other (as marked by the blue arrows) at 7.5 min. Moreover, we found that in each of these two pairs the two BPs were located so close to each other and crossed each other's path (Figure 7). This might indicate the entanglement of the flux tubes. These observed characteristics are quite consistent with the process of magnetic reconnection near the photosphere resulting in flux cancelation, mutual collision, and disappearance of two magnetic flux elements of opposite polarity.

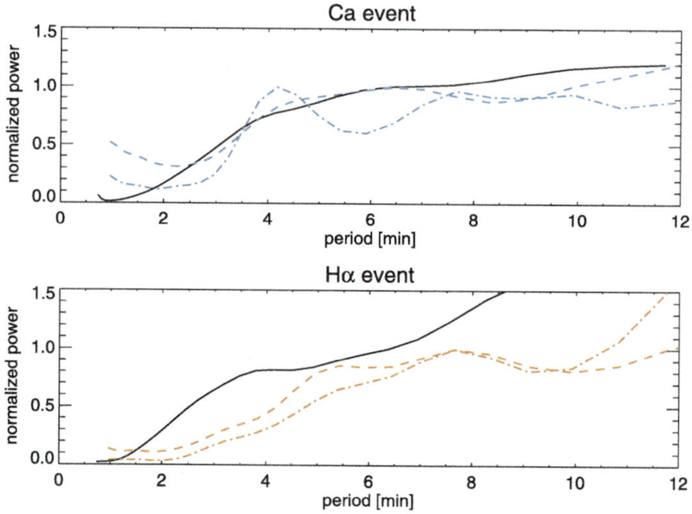

Figure 8 Top: Power spectra at the site of the Ca II event. The solid black curve represents the oscillatory power of TiO data, the dash-dotted blue curve shows the oscillatory power of Ca II intensity, and the blue dashed curve shows the power of Ca II velocity. Bottom: The same for the Hα event. The oscillatory power of TiO intensity is represented with the solid black curve, the oscillatory power of Hα intensity is shown with the dash-dotted orange curve, and the oscillatory power of Hα velocity is shown with the dashed orange curve.

3.4. Wavelet Analysis

We performed the wavelet analysis to detect waves connected with these events. For the TiO data set we analyzed intensity images, while for the Hα and Ca II data sets, we analyzed intensity and velocity data. We focus on the results of wavelet analysis at the locations and times of our events (Figure 8).

The figure shows that both events accompanied oscillation phenomena that may be considered as an observational manifestation of waves. The power spectrum of Ca II intensity in the Ca II event shows a strong peak at a period of ≈ 4 min. Two smaller peaks also appear at ≈ 7 and ≈ 10 min. In comparison, the power spectrum of TiO intensity data monotonically increases with period. The power spectrum of Hα intensity in the Hα event shows the most prominent peak at ≈ 7.5 min, and weak enhancement at ≈ 3 min. The power spectrum of velocity data has small peaks at 5 and 7.5 min. The power spectrum of TiO intensity increases with period like in the Ca II event, but has a small peak at ≈ 3.5 min.

3.5. Phase Differences

We analyzed phase differences between intensity and velocity (I-V) for the Hα and Ca II lines (Figure 9). The phase difference is set to zero if the peak brightening in the intensity signal occurs at the same time as the peak blueshift. According to a theory of the propagation of acoustic waves (Lites and Chipman, 1979), the upward velocity of an upward propagating acoustic wave should be in phase with the enhancements in pressure, temperature, density, and hence intensity. Thus, zero phase between intensity and velocity signal indicates upward propagation in an acoustic wave.

Figure 9 shows a complicated nature of the observed oscillations. The simplest wave trains are observed in the blue wings of both lines (Figures 9C and F), which may indicate

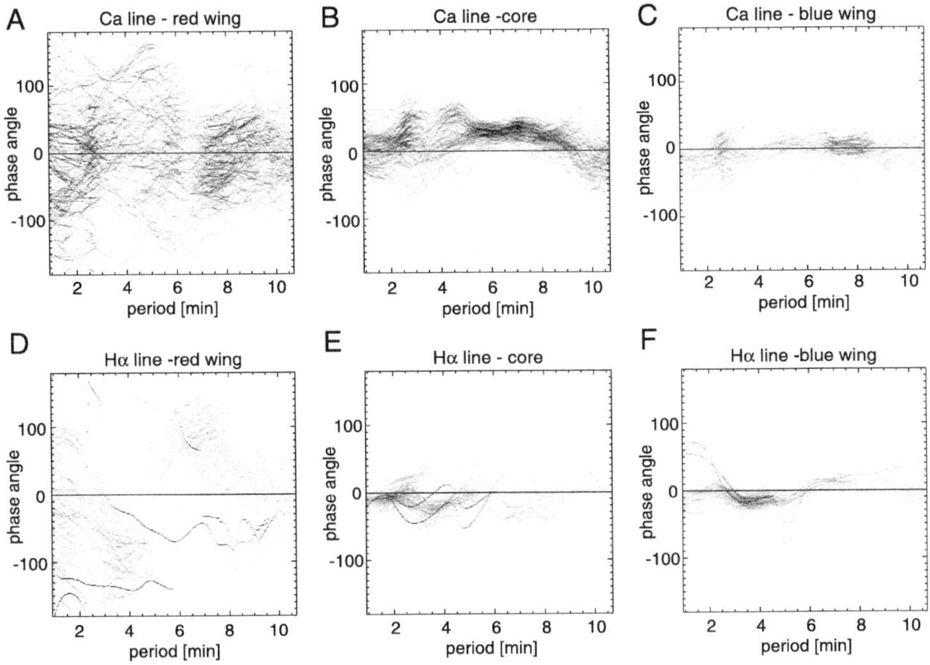

Figure 9 Top: Phase differences between intensity and velocity signals from the Ca II line. Panels A, B, and C are for red wing, line core, and blue wing, respectively. Bottom: Phase differences between intensity and velocity signals from the Hα line. Panels D, E, and F are for the red wing, line core, and blue wing, respectively.

acoustic wave trains. It appears, however, that the Ca II and Hα lines rather reflect the superposition of various wave modes. The most complicated wave trains are in the red wings of both lines (Figures 9A and D) where we see a plethora of different waves. The line cores (Figures 9B and E) also show mixture of different wave modes. These observed mixtures of phase differences suggest that in fact there may be more than one wave train, as predicted by Bogdan *et al.* (2003). Contributions from non-acoustics waves are also possible.

In Figure 10 we have presented phase differences between the Ca II line and the Hα line. The positive phase difference means that the Hα line signal follows that of the Ca II line. Since the Hα line is formed at a little higher level (1500 km above $\tau_{500\ \mathrm{nm}} = 1$; Vernazza, Avrett, and Loeser, 1981) than the Ca II line (1200 km; Cauzzi *et al.*, 2008), the phase difference between these lines is expected to be small and positive for upward propagating waves. But the V-V signals for either event do not show any clear signature of such expectation, only indicating a complex mixture of wave trains propagating in any direction with any speed (Figures 10B and D). Intensity phase differences in the Hα event show a similar situation (Figure 10C), namely a mixture of phase angles as V-V signals. Only the intensity phase differences found in the Ca II event show clear indication of positive phase differences in the period range from 4 to 6.5 min (Figure 10A). The detected positive phase difference means that the intensity signal first appears in the core of the Ca II line and then in the core of the Hα line. However, the difference between the formation heights of these two spectral lines is too small to explain these large phase differences in terms of simple upward propagating waves.

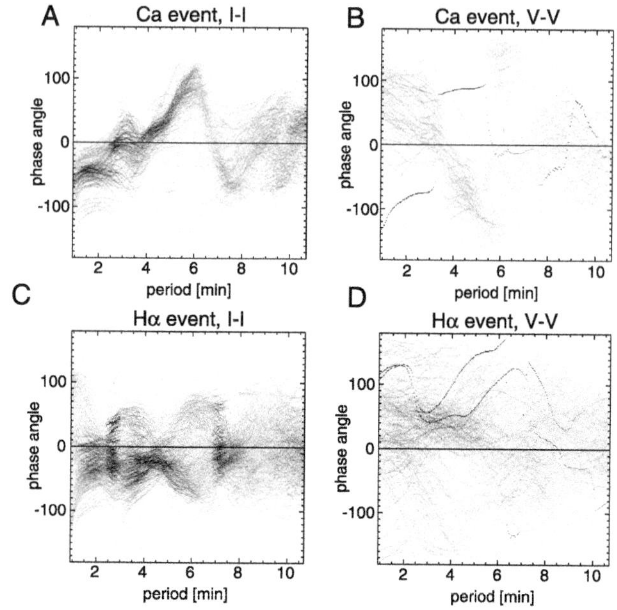

Figure 10 Phase differences for the Ca II (top) and Hα (bottom) events. Phase differences between Ca II and Hα lines are derived from intensity signals (left) and velocity signals (right), respectively. Positive phase difference means that a signal first appears in the Ca II line and then in the Hα line.

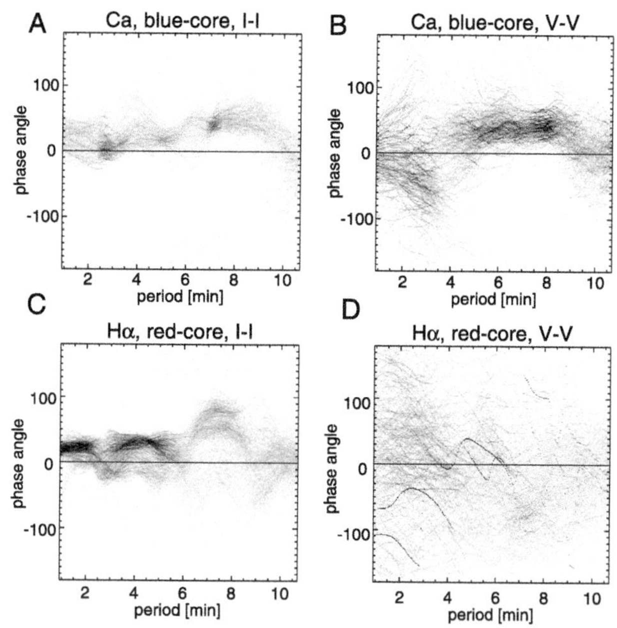

Figure 11 Phase differences for the Ca II (top) and Hα (bottom) events. Phase differences are derived from (A) blue wing and line core intensities of Ca II, (B) blue wing and line core velocities of Ca II, (C) red wing and line core intensities of Hα, and (D) red wing and line core velocities of Hα, respectively. Positive phase difference means that a signal first appears in the wing and then in the core of the line.

Since we detected the upflow from the Ca II line and the downflow from the Hα line in the velocity maps, we also examined any propagation of signals from the blue wing to the line core of the Ca II line and from the red wing to the line core of the Hα line (Figure 11). For the upflow event we have a clear propagation signature from the blue wing to the core in intensity and velocity signals (Figures 11A and B). The intensity signals show propagation

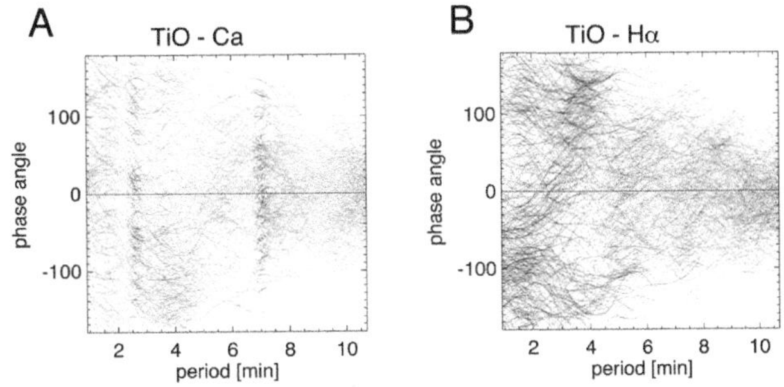

Figure 12 Phase differences between (A) TiO and Ca II intensities and (B) TiO and Hα intensities, respectively. Positive phase difference means that a signal first appears in the TiO line and then in the Ca II or Hα line.

from the wing to the core approximately over the whole period range (Figure 11A), while the velocity signals show clear propagation only in a period range of 5 to 9 min (Figure 11B).

The downflow event in Hα shows a different situation. In the intensity signals we see propagation from the red wing to the line core (Figure 11C) while the velocity signals show a chaotic mixture of waves moving in every direction with every speed (Figure 11D).

Previous researches suggested that Ca II brightening is caused by waves emitted from the photosphere. We tested this connection between the chromospheric events (Ca II and Hα) and photospheric waves (TiO) using their phase differences (Figure 12). In Figure 12 positive phase difference means that a signal first appears in TiO and then in Ca II or Hα. It is found that waves propagate simultaneously in upward and downward directions with various phase differences, and no clear trends are visible. Namely, the observed waves do not show any direct connection between the observed chromospheric events and photospheric dynamics below them.

Note that the phase differences shown in Figures 9, 10, 11, and 12 are weighted with power of the observed waves, making more powerful waves appear darker in the plots. Figure 12A shows that approximately all phase angles have some power, but we see a darker pattern appearing at periods ≈ 3 and ≈ 7 min. The concentrations of power near ≈ 3 min may indicate dominance of downward propagation (negative phase difference) while those near ≈ 7 min propagate in both directions, upward and downward. Figure 12B also shows concentrations of power in the Hα downflow event. In this case we see concentrations of power in the upward propagation with various phase angles in a period of ≈ 3.5 min and also in the downward propagation in the period range from 0.6 to ≈ 3.5 min.

3.6. Fourier–Hilbert Analysis

We applied the Fourier–Hilbert method to the TiO data to learn more about photospheric waves and their propagation direction. To make results comparable to the results of previous work, the amplitude and phase were integrated over the whole observing time, and were normalized with respect to their maximum value (Lites and Chipman, 1979). Then we removed all waves with amplitudes lower than 5 % of the maximum amplitude. In Figure 13 the amplitudes and phases thus derived are shown. The phase differences were presented in a weighted diagram, where the weight was applied by utilizing a cross-power amplitude $\sqrt{P_1 P_2}$. Here positive phase means upward propagating waves. For the Ca II event, where the BPs were located at around $(1'', 1.2'')$, we see intense downward propagation of observed waves (Figure 13C). For the Hα event the situation is different; in the area where the intergranular lane was broadened we see strong upward propagation of the waves (Figure 13D).

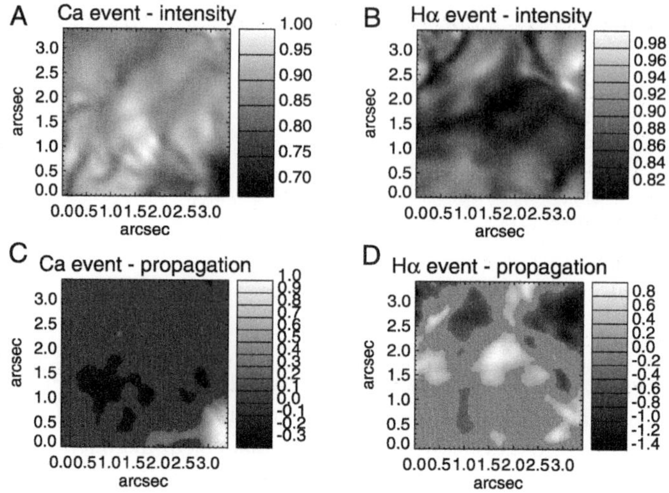

Figure 13 Normalized amplitudes (top) and phases (bottom) derived from the Fourier–Hilbert method applied to the TiO data. The left and right panels are for the Ca II and Hα events, respectively. The phase and amplitude were integrated over the observing time and normalized with respect to their maximum positive values. Then we removed all waves with amplitudes lower than 5 % of the maximum amplitude.

4. Discussion

We observed a Ca II brightening in one part of the FOV (Figure 3) with a strong upflow. In its vicinity we detected simultaneous strong downflow in the Hα line (Figure 2). The photosphere below the brightening area showed indications of possible flux tube entanglement (Figure 7), while at the location of the downflow we observed enhanced intergranular lanes (Figure 4). At a first glance it seems that the behavior of the photosphere corresponded to the observed events in the chromosphere.

Researchers in the past found that Ca II brightening events are related to magnetic elements (Sivaraman and Livingston, 1982; Sivaraman *et al.*, 2000), while brightening events in Hα or Ellerman bombs are not so, usually (Nindos and Zirin, 1998). Our findings agree with this work. We observed a Ca II brightening event above a cluster of BPs, without the accompanying Ellerman bombs (Figures 3 and 7). The location and intensity changes of the observed Ca II brightening fit the results of previous studies (Sivaraman and Livingston, 1982; Sivaraman *et al.*, 2000).

Kariyappa, Narayanan, and Dame (2005) observed the 3 min period waves in the Ca II brightening and speculated that they are due to chromospheric oscillations. Our work is in agreement with this speculation. Carlsson and Stein (1992, 1997) simulated the generation of Ca II brightening with acoustic shocks. However, we did not see this connection in our event.

Bogdan *et al.* (2003) pointed out that the location of the magnetic canopy and the location of the wave sources with their dominant state of polarization (radial vs. transverse driving motions) are important factors in considering the properties of propagating waves. The existence of a pore inside the FOV shows that parts of the canopy touched the photosphere inside the FOV, setting the canopy lower than in a typical quiet sun. Moreover, our FOV was close to the active region, AR 11087, which complicated the situation further. Our two spectral lines, Ca II and Hα, are formed in the height range of the canopy itself. Since we studied the wings and the centers of the lines, we can state that FISS data sets covered significant portions of the canopy, but it is hard to say which portions exactly they did. Even from the simple quiet sun model by Vernazza, Avrett, and Loeser (1981) we see that the Ca II and Hα lines cover a wide range of atmospheric heights. Since inside the canopy the plasma conditions vary from one flux tube to the next it is hard to establish the exact range of spectral line formation. Thus, the closest we can state with certainty is that the Ca II and Hα lines are formed in the canopy. With the TiO-band observations the situation is clearer, data give information on the low photosphere.

We started our analysis with the assumption that the sources are located in the photosphere in the intergranular lanes, which is a typical situation for photospheric waves. The assumption was partially correct, as seen in Figure 13, but it did not provide the full answer since the phase differences between photospheric (TiO band) and chromospheric data did not agree with this picture (Figure 12).

Instead, the waves observed in the event are complex and point to a more than one oscillatory source being involved. The previous work established the existence of propagating waves in the flux tubes (DePontieu, Erdélÿi, and James, 2004; Jess *et al.*, 2007; Andić *et al.*, 2010). Thus, one could expect to observe the wave propagation between photospheric and chromospheric layers especially when we observe along the axis of a flux tube. According to Bogdan *et al.* (2003) the propagation of the waves could be complex especially when the magnetic canopy is located in a part of the propagating path. Bogdan *et al.* (2003) found that the canopy area in the propagating path shows a very complex wave behavior in which fast waves are expected to propagate in a wide range of directions, especially in the direction close to the tangent of the magnetic loops. The complexity of our phase difference spectra shown in Figure 12 could be explained with the presence of a similar condition somewhere between the formation heights of the TiO band and the Ca II and Hα lines.

Most of the observed waves in Ca II and Hα during these two events show that they are not pure acoustic propagating waves (Figure 9), since for the acoustic waves one expect the brightening (intensity increase) to be in phase with the velocity of upflow (Lites and Chipman, 1979). The nature of the waves also confirms that they originate from additional sources, not only from the previously established photospheric sources which are thought to produce mainly acoustic waves (Nordlund and Stein, 2001; Khomenko, Collados, and Felipe, 2008; Andić *et al.*, 2010). Thus we can conclude that the upflow event is not connected with the waves created in the photosphere directly below it, but has a different origin.

From Figure 11 we see that there is a propagation of signals from the wings to the core of the line for the Ca II brightening event. But there was no such propagation between the photosphere and the Ca II formation height (Figure 12). Thus we might speculate that those propagating waves observed in the Ca II line most likely originate from the same mechanism as the observed upflow (Figure 3), and that complex propagation characteristics seen between the formation layers of the TiO band and of the Ca II line is partly caused by the same source emitting waves in all directions.

On the other hand, it is possible that waves observed in the chromosphere come from somewhere else via one of the many flux tubes that are intervened in the area. Unfortunately, due to our limited data sets without magnetic field information, we are not able to reconstruct the possible shapes and positions of the flux tubes, and consequently other possible candidates for the oscillatory sources in the photosphere are not excluded.

The downflow event observed in Hα was simultaneous with the upflow observed in Ca II, which made us believe that due to the simultaneous nature of the events and their closeness they are connected. However, we found no proof that those events are connected. The simultaneous changes in intensity of both events are most probably coincidental. The event itself shows strong downflow of plasma (Figure 2) that coincided with the persistent presence of the intergranular lanes in the photosphere below the event (Figure 4). This indicates that the plasma downflow noted in Hα might have continued down to the photosphere.

The properties of waves are more complicated for the downflow event. Figure 13 shows that on the location of the downflow we have an emission of the upward propagating waves. This is in agreement with the previous studies (Andić, 2007b; and references therein). However, by examining the direction of wave propagation between the photosphere and the Hα formation height we encountered a complicated situation (Figure 12) that the wave trains propagated in both directions with a multitude of phase angles. These results are similar to the complexities of wave propagation demonstrated by Bogdan *et al.* (2003). This hypothesis is supported by the assumed location of the canopy in our data set. Figure 9D, E, and F demonstrate that the waves observed in Hα are a mixture of various wave modes. Figure 11D shows that the complexity of wave propagation continued to the Hα line formation height. Figure 11C is the only phase spectrum that clearly shows a propagation from the red wing to the core of the Hα line. Even in this case a general impression is that the wave trains in the plasma in the downflow event are complex with multiple wave sources and wave modes and thus require more detailed data sets combined with theoretical modeling to fully understand what is going on with the waves during the plasma downflow of this kind.

5. Conclusions

We observed the Ca II brightening event that coincided with the plasma upflow and it was located above the magnetic elements in the photosphere. However, we could not observe the connection between the waves in the photosphere and those in the Ca II brightening area. We speculate that the waves in the brightening area are alternatively caused by magnetic reconnection that is most likely to have caused brightening itself and the plasma upflow. As Bogdan *et al.* (2003) demonstrated, the behavior of waves is complex in the canopy configuration, and observations of the photosphere and the lower chromosphere with instruments that allow 2D spectropolarimetry would be beneficial to investigate complicated processes going on in this height range.

In the case of the plasma downflow event, we found that upward propagating waves are created in the intergranular lanes. However, the propagation of the waves became complicated when the chromosphere is included into the analysis. Although it is possible that the plasma downflow continued from the Hα line formation height all the way down to the low photosphere, the nature of the propagating waves is not perfectly clear. It seems that multiple oscillatory sources and/or sites were involved where the waves might be reflected from the magnetic canopy and might change modes (Bogdan *et al.*, 2003). Unfortunately, this last part is only speculation, since the places of reflection and additional sources could not be identified in our data set.

Acknowledgements The authors are thankful to the anonymous referee for helpful comments. AA is thankful to J. Varsik and R. Fear for the help with the observations. K. Ahn and W. Cao acknowledge the support of the US NSF AGS-0847126. The work of J. Chae, H. Park, and H. Yang was supported by the National Research Foundation of Korea (KRF-2008-220-C00022), and that of Y.D. Park, by the Development of Korean Space Weather Center, a project of KASI.

References

Ahn, K., Chae, J., Park, H.M., Nah, J., Park, Y.D., Jang, B.H., Moon, Y.J.: 2008, *J. Korean Astron. Soc.* **41**, 39.

Andjic, A.: 2006, *Serb. Astron. J.* **172**, 27.
Andić, A.: 2007a, *Solar Phys.* **242**, 9.
Andić, A.: 2007b, *Solar Phys.* **243**, 131.
Andić, A., Goode, P.R., Chae, J., Cao, W., Ahn, K., Yurchyshyn, V., Abramenko, V.: 2010, *Astrophys. J. Lett.* **717**, L79.
Bloomfield, D.S., McAteer, R.T.J., Lites, B.W., Judge, P.G., Mathioudakis, M., Keenan, F.P.: 2004, *Astrophys. J.* **617**, 623.
Bloomfield, D.S., McAteer, R.T.J., Mathioudakis, M., Keenan, F.P.: 2006, *Astrophys. J.* **652**, 812.
Bodo, G., Kalkofen, W., Massaglia, S., Rossi, P.: 2001, *Astron. Astrophys.* **370**, 1088.
Bogdan, T.J., Carlsson, M., Hansteen, V., McMurry, A., Rosenthal, C.S., Jonson, M., Petty-Powell, S., Zita, E.J., Stein, R.F., McIntosh, S.W.: Nordlund, Å.: 2003, *Astrophys. J.* **599**, 626.
Cao, W., Gorceix, N., Coulter, R., Woeger, F., Ahn, K., Shumko, S., Varsik, J., Coulter, A., Goode, P.R.: 2010, In: McLean, I.S., Ramsay, S.K., Takami, H. (eds.) *Ground-Based and Airborne Instrumentation for Astronomy III*, *Proc. SPIE* **7735**, 77355V.
Cauzzi, G., Reardon, K.P., Uitenbroek, H., Cavallini, F., Falchi, A., Falciani, R., Janssen, K., Rimmele, T., Vecchio, A., Wöger, F.: 2008, *Astron. Astrophys.* **480**, 515.
Chae, J., Sakurai, T.: 2008, *Astrophys. J.* **689**, 593.
Chae, J., Goode, P.R., Ahn, K., Yurchycsyn, V., Abramenko, V., Andić, A., Cao, W., Park, Y.D.: 2010, *Astrophys. J. Lett.* **713**, L6.
Carlsson, M., Stein, R.R.: 1992, *Astrophys. J. Lett.* **397**, L59.
Carlsson, M., Stein, R.F.: 1997, *Astrophys. J.* **481**, 500.
De Moortel, I., Galsgaard, K.: 2006, *Astron. Astrophys.* **451**, 1101.
DePontieu, B., Erdélÿi, R., James, S.P.: 2004, *Nature* **430**, 536.
Ellerman, F.: 1917, *Astrophys. J.* **46**, 298.
Goode, P.R., Yurchyshyn, V., Cao, W., Abramenko, V., Andić, A., Ahn, K., Chae, J.: 2010, *Astrophys. J. Lett.* **714**, L31.
Jain, R., Haber, D.: 2002, *Astron. Astrophys.* **387**, 1092.
Janssen, K.: 2003, *Struktur und Dynamik kleinskaliger Magnetfelder der Sonnenatmosphaere*, Copernikus GmbH, Kaltenburg-Lindau, 123.
Jess, D.B., McAteer, R.T.J., Mathioudakis, M., Keenan, F.P., Andić, A., Bloomfield, D.S.: 2007, *Astron. Astrophys.* **476**, 971.
Kalkofen, W.: 1999, In: Schmieder, B., Hofmann, A., Staude, J. (eds.) *Third Advances in Solar Physics Euroconference: Magnetic Fields and Oscillations*, *ASP Conf. Ser.* **184**, 227.
Kalkofen, W.: 1997, *Astrophys. J. Lett.* **486**, L145.
Kariyappa, R., Satya Narayanan, A., Damé, L.: 2005, *Bull. Astron. Soc. India* **33**, 19.
Khomenko, E., Collados, M., Felipe, T.: 2008, *Solar Phys.* **251**, 589.
Leibacher, J., Gouttebroze, P., Stein, R.F.: 1982, *Astrophys. J.* **258**, 393.
Lites, B.W., Chipman, E.G.: 1979, *Astrophys. J.* **231**, 570.
Musielak, Z.E., Ulmschneider, P.: 2003, *Astron. Astrophys.* **406**, 725.
Nindos, A., Zirin, H.: 1998, *Solar Phys.* **182**, 381.
Noble, M.W., Musielak, Z.E., Ulmschneider, P.: 2003, *Astron. Astrophys.* **409**, 1085.
Nordlund, Å., Stein, R.F.: 2001, *Astrophys. J.* **546**, 576.
Reardon, K.P., Leperti, F., Carbone, V., Vecchio, A.: 2008, *Astrophys. J. Lett.* **683**, L207.
Rosenthal, C.S., Bogdan, T.J., Carlsson, M., Dorch, S.B.F., Hansteen, V., McIntosh, S.W., McMurry, A., Nordlund, Å., Stein, R.F.: 2002, *Astrophys. J.* **564**, 508.
Shelyag, S., Zharkov, S., Fedun, V., Erdélyi, R., Thompson, M.J.: 2009, *Astron. Astrophys.* **501**, 735.
Sivaraman, K.R., Livingston, W.C.: 1982, *Solar Phys.* **80**, 227.
Sivaraman, K.R., Gupta, S.S., Livingston, W.C., Damé, L., Kalkofen, W., Keller, C.U., Smartt, R., Hasan, S.S.: 2000, *Astron. Astrophys.* **363**, 279.
Stebbins, R., Goode, P.R.: 1987, *Solar Phys.* **110**, 237.
Torrence, C., Compo, G.P.: 1998, *Bull. Am. Meteorol. Soc.* **79**, 61.
Vernazza, J.E., Avrett, E.H., Loeser, R.: 1981, *Astrophys. J. Suppl.* **45**, 635.
Volkmer, R., Kneer, F., Bendlin, C.: 1995, *Astron. Astrophys.* **304**, 1.
von der Lühe, O.: 1993, *Astron. Astrophys.* **268**, 347.
White, O.R., Cha, M.Y.: 1973, *Solar Phys.* **31**, 23.
Wöger, F., von der Lühe, O., Reardon, K.: 2008, *Astron. Astrophys.* **488**, 375.

DOI 10.1007/978-3-319-12123-9_5
Reprinted from *Solar Physics* Journal, DOI 10.1007/s11207-013-0286-8

Chromospheric Sunspot Oscillations in Hα and Ca II 8542 Å

Ram Ajor Maurya · Jongchul Chae · Hyungmin Park · Heesu Yang · Donguk Song · Kyuhyoun Cho

Received: 30 June 2012 / Accepted: 24 March 2013 / Published online: 23 April 2013
© Springer Science+Business Media Dordrecht 2013

Abstract We study chromospheric oscillations including umbral flashes and running penumbral waves in a sunspot of active region NOAA 11242 using scanning spectroscopy in Hα and Ca II 8542 Å with the *Fast Imaging Solar Spectrograph* (FISS) at the 1.6 meter *New Solar Telescope* at the Big Bear Solar Observatory. A bisector method is applied to spectral observations to construct chromospheric Doppler-velocity maps. Temporal-sequence analysis of these shows enhanced high-frequency oscillations inside the sunspot umbra in both lines. Their peak frequency gradually decreases outward from the umbra. The oscillation power is found to be associated with magnetic-field strength and inclination, with different relationships in different frequency bands.

Keywords Oscillations, solar · Chromosphere, active · Sunspots, penumbra · Sunspots, umbra · Waves, propagation

1. Introduction

Umbral flashes (UFs) and running penumbral waves (RPWs) are long-known sunspot-oscillation phenomena that have been studied extensively. The first is more three-minute, the latter more five-minute in character.

UFs were first discovered by Beckers and Tallant (1969) in Ca II H and K filtergrams and spectrograms observations of a sunspot. These authors suggested that the UFs may be caused by the magneto-acoustic wave phenomenon. Following a suggestion by Beckers and Tallant (1969), Havnes (1970) theoretically showed that UF brightening occurs in the local umbral gas during the compressional phase of a magneto-acoustic wave; the compression causes a rise in the temperature and a corresponding increase in number density of Ca II atoms. Recently, it has been confirmed by Bard and Carlsson (2010) from NLTE simulations of

Initial Results from FISS
Guest Editor: Jongchul Chae

R.A. Maurya (✉) · J. Chae · H. Park · H. Yang · D. Song · K. Cho
Department of Physics and Astronomy, Seoul National University, Seoul 151-747, Republic of Korea
e-mail: ramajor@astro.snu.ac.kr

the Ca II H line. They concluded that UFs result from increased emission of the local solar material during the passage of acoustic waves originating in the photosphere and steepening to shock in the chromosphere.

RPWs were studied first by Zirin and Stein (1972) and Giovanelli (1972), independently, in Hα filtergrams; they believed the RPWs to be acoustic type and Alfvén type, respectively. A reasonably detailed theory of RPWs in terms of magneto-atmospheric waves was given by Nye and Thomas (1974, 1976). From a simple model of the penumbral atmosphere with a purely horizontal magnetic field, they concluded that the RPWs are gravity-modified fast magneto-acoustic waves that are vertically trapped at the photospheric levels. The waves are more nearly acoustic type at the low levels (convection zone), while they are more nearly of Alfvén type at higher levels (photosphere and low chromosphere). The maximum vertical velocity occurs in the chromosphere, but the density is very low there and most of the wave energy lies in the photosphere and sub-photosphere, where the vertical velocity is lower but the density is much higher.

In recent observations, RPWs appear as a sort of continuation of some of the flash waves. The main difference between the UFs and the RPWs is that the UFs are best seen in Ca II H and K lines, while RPWs are best seen in Hα Dopplergrams. The flashes are present only when the oscillation amplitude is sufficiently large ($\geq 5\ \mathrm{km\,s^{-1}}$), but oscillatory motions are always present in nearly every umbra (Moore, 1981). Furthermore, UFs and RPWs have been interpreted as different manifestations of the same oscillatory phenomenon in a sunspot, combining upward-shock propagation with coherent wave spreading over the entire spot (Zhugzhda and Dzhalilov, 1984; Rouppe van der Voort *et al.*, 2003; Bloomfield, Lagg, and Solanki, 2007).

However, it is still unclear what determines the characteristic properties of the oscillations in sunspot chromospheres. How do the waves that are associated with these oscillations propagate in the magnetized atmosphere? What is the spatial variation of frequency across the sunspot and in features of different spatial scales? The oscillations associated with waves are important to study since they carry some information on the properties of the region from which they originate and through which they propagate. Furthermore, the rapidly evolving field of local helioseismology uses the oscillation properties to investigate the sub-surface structure and dynamics.

UFs and RPWs appear more clearly in Dopplergrams than in intensities. In many earlier studies, the chromospheric oscillations were analyzed by constructing Dopplergrams from filtergrams at fixed wavelengths in the red and blue wings. In this article, simultaneous spectral observations in the Hα and Ca II 8542 Å lines have been used to construct Doppler-velocity maps (hereafter Doppler maps) from the bisectors of these two lines.

The rest of the article is organized as follows: We describe the observational data in Section 2 and present the methods of analysis in Section 3. Results and discussions are given in Section 4. Finally, Section 5 is devoted to the summary and conclusions.

2. Observations

We observed Active Region (AR) NOAA 11242 on 30 June 2011. It was located at heliographic latitude 17°N and longitude 29°W, and consisted of a well-developed sunspot of southern polarity with mean magnetic-field strength 1200 gauss. The sunspot was surrounded by a number of small magnetic fragments of northern polarity. The AR was well observed by the *Fast Imaging Solar Spectrograph* (FISS) instrument (Chae *et al.*, 2012), and the *Helioseismic and Magnetic Imager* (HMI: Schou *et al.*, 2012) onboard the *Solar Dynamics Observatory* (SDO).

FISS is a slit spectrograph taking spectra with a rapid-scan capability. It observes the solar chromosphere simultaneously in two spectral bands centered around the lines Hα and Ca II 8542 Å (hereafter Ca II) with spectral resolutions of 0.045 Å and 0.064 Å, respectively. The pixel resolution at both wavelengths is $\approx 0.16''$ pixel^{-1}.

The FISS data cube, with a field of view of 24″ of scan width (total number of scans 150) and slit length of 41″, covers the entire sunspot of AR NOAA 11242. The observations were taken under relatively good seeing conditions during 17:51:38 – 18:55:42 UT, except for a gap during 18:04:45 – 18:18:10 UT, with the scan step sampling, timing, and cadence of 27 seconds, 130 milliseconds, and 30 seconds, respectively, in both spectral bands.

In order to study the association between chromospheric oscillations and magnetic fields, we have used the HMI vector magnetograms: field strength, azimuth angle, and angle of inclination. The angle of inclination is measured from the line-of-sight and azimuth from the $+y$-direction of the heliocentric coordinates (x, y).

3. Data Reduction

Out of 105 cubes for each of Hα and Ca II, seven were discarded, being of low quality due to failure of adaptive optics (AO) locking. The data are processed in two main steps. In the first step, we correct the data for bias, dark, flat, and slit patterns. In the second step the wavelength scales are calibrated. Finally the processed data are compressed using principal component analysis (PCA: Pearson, 1901), which is useful especially when spectral profiles are similar to one another (Rees *et al.*, 2000). PCA compression suppresses the noise to a substantial amount without much loss of information. More details of the processing and PCA compression can be found in Chae *et al.* (2012).

3.1. Alignment

Image rotation at the Coudé focus, telescope guiding errors, errors in slit positioning, and seeing cause shifts between successive scans. We used the far-wing parts of the scans as references. The successive scans were first derotated and then aligned by cross-correlation to the reference scan, and the Hα and Ca II scans were also aligned by cross-correlation. The error due to slit positioning is found to be $\pm 0.16''$. Typical values of the net shifts in scan x- and y-directions were found to be eight pixels and two pixels, respectively. The FISS data were aligned with the HMI images, by manual feature and pattern matching, into solar (x, y) coordinates. Figure 1 shows a mosaic of the aligned FISS and HMI images of AR NOAA 11242 on 30 June 2011.

3.2. Timing Correction

Both the image rotation and the image shifts between successive scans that were corrected by derotation and alignment suggest that the timing per pixel does not have a fixed cadence. We determined the actual observing times per pixel from the alignment vectors.

Figure 2 shows the temporal variation in cadence $[\delta(\Delta t)]$ for three locations (1, 2, and 3, marked with circles in Figure 1b). There is a data gap between 18:04:45 – 18:18:10 UT, and also the images have large shift and rotation. In addition, $\delta(\Delta t)$ varies in the range ± 5 seconds, for a total observation period of 63 minutes. The deviation $\delta(\Delta t)$ for the point 1 (3) becomes more negative (positive) with time, which suggests that the pixel corresponding to this position was observed earlier (later). Note that we have used the data sets starting only from the time 18:18:10 UT, for the oscillation studies. Also, the seven bad-AO scans were excluded.

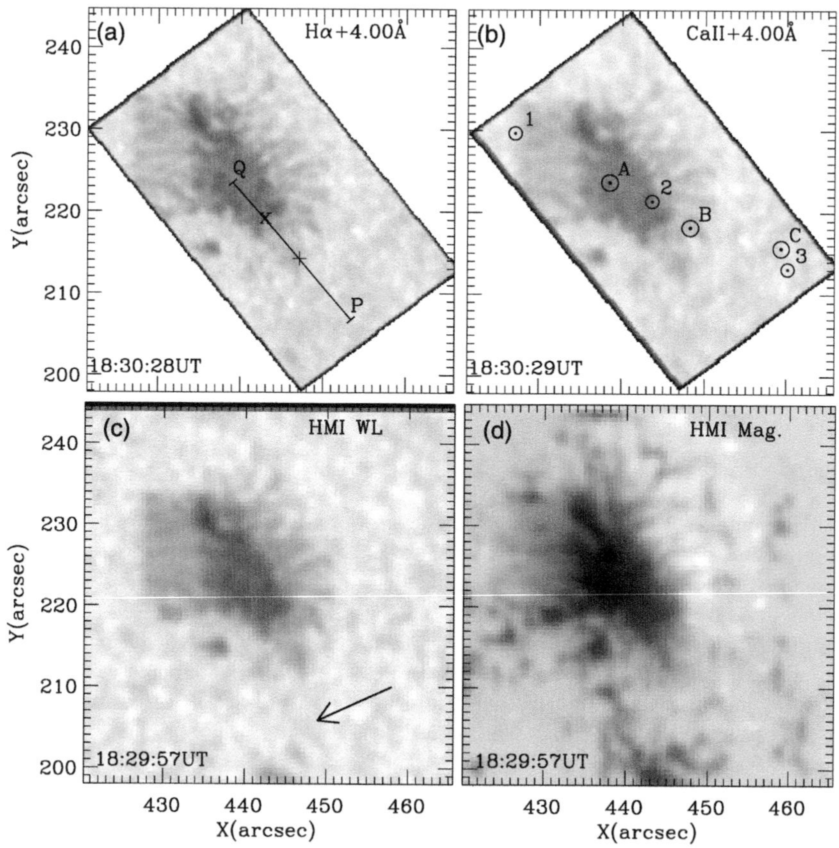

Figure 1 Mosaic of images of AR NOAA 11242 on 30 June 2012 in different wavelengths. The upper panels show the raster images at wavelengths 4.0 Å from Hα (a) and Ca II (b) lines centers. The line PQ (a) marks reference locations for the analysis. The locations marked by the cross (×) and plus (+) symbols represent the approximate boundaries of the umbra and the penumbra, respectively. Small circles labeled with integer numbers (b) mark the locations for which time information is shown in Figure 2, while larger circles labeled with A, B, and C represent three locations as in Figure 4. The lower panels show the HMI continuum (c) and magnetogram (d) images observed near the time of the FISS observation (upper panels). The arrow in panel (c) denotes the direction of the solar disk center.

3.3. Bisector Measurement, Doppler Maps, and Power Maps

We compute the Doppler velocity using the bisectors (Gray, 1976; Dravins, Lindegren, and Nordlund, 1981) of spectral-line profiles. The locus of bisectors represents the asymmetry of the spectral line around the line center. For the bisector for a given spatial location, one may sample the profile asymmetry at a given intensity level (Keil and Yackovich, 1981; Cavallini *et al.*, 1986), at a given separation from the nominal line-center wavelength or the per-pixel measured minimum wavelength (Bhatnagar, Livingston, and Harvey, 1972; von Uexküll, Kneer, and Mattig, 1983; Tziotziou, Tsiropoula, and Mein, 2002; Tziotziou *et al.*, 2007), and lambdameter measurement at given chord length (Slaughter and Wilson, 1972; Stebbins and Goode, 1987). We prefer to measure Doppler velocities with the lambdameter method.

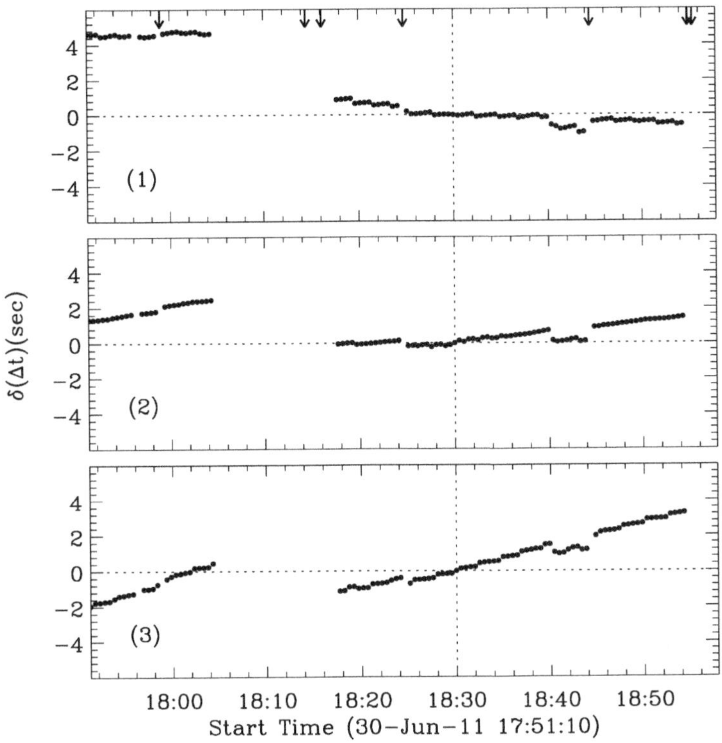

Figure 2 Timing shifts due to image rotation and scanner shifts at the three locations marked in Figure 1b. Vertical dotted line specifies the reference image. Arrows in the upper panel mark the locations of bad-AO scans.

Of course, the lambdameter method has a problem, along with the first two methods, in the case when the Doppler width is changing, *e.g.* spectral lines are wider in the sunspot umbra than in a quiet region. There is a possibility of width variation with time for a given location due to energetic activities. However, it is difficult to remove the width-related variation in the Doppler velocity for every pixel in the time series. To minimize this effect, we have chosen an optimum value of the bisector chord, $\delta\lambda = 0.35$ Å (0.15 Å) for Hα (Ca II). These chords were applied to all the spatial pixels to compute the Doppler maps (see Figure 4, bottom row). To average over remaining image distortions due to seeing, guiding errors, *etc.*, we applied spatial 3×3 boxcar smoothing. For the power analysis of Doppler maps, we prefer not to interpolate the data to equal temporal sampling but instead used the Lomb–Scargle periodogram (LSP) technique (Lomb, 1976; Scargle, 1982).

4. Results and Discussion

Figure 3 shows time-averaged intensity profiles and bisectors of Hα (a) and Ca II (b) lines for location B (see Figure 1b and Figure 4).

The bisectors for the Hα line are "C" shaped while for the Ca II line they are inverse "C", and represent blue and red asymmetry, respectively, in the line profiles. Uitenbroek (2006)

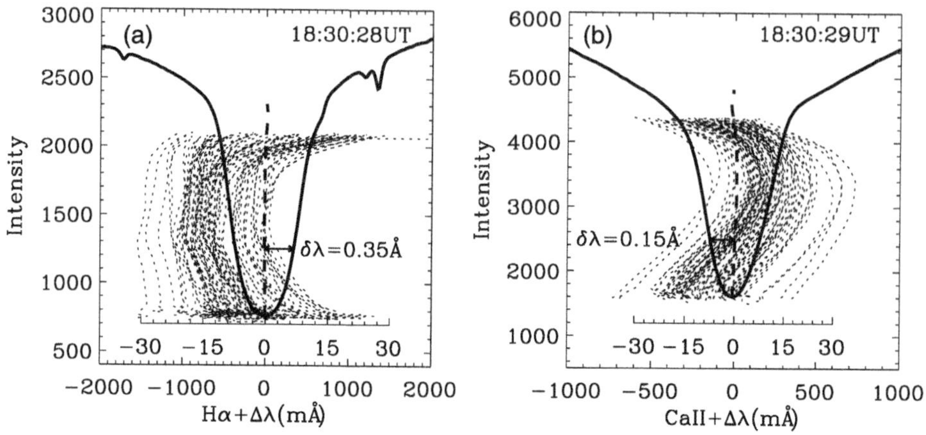

Figure 3 Time-average intensity profiles (thick-solid) and bisectors (thick-dashed) for the location B (see Figure 4). Thin-dashed curves, plotted with respect to x-axes drawn near lines' cores, represent bisectors on a magnified scale (50 times for Hα and 20 times for Ca II) sampled at the same location at different times. The horizontal arrow marks the bisector chord $\delta\lambda = 0.35$ Å (0.15 Å) for Hα (Ca II).

has also reported inverse "C" shaped bisectors and a strong red asymmetry in the Ca II line. The amount of curvature, or, equivalently, the velocity covered by the bisector is a measure of asymmetric flows.

Figure 4 shows samples of chromospheric Doppler maps of the Hα and Ca II lines, respectively, constructed at bisector chords $\delta\lambda = 0.35$ Å and $\delta\lambda = 0.15$ Å. The Doppler velocities in these maps range ± 6 km s^{-1}, where negative/positive velocity shows upward/downward plasma motions. Patterns associated with running penumbral waves, in Hα and Ca II, can be seen in the Doppler maps. The overall red–blue asymmetry of the Doppler maps is caused by off-disk-center viewing. The blue in the limbward part of the penumbra indicates the line-of-sight component of the reverse Evershed flow. It is less evident on the center-side, although filaments there would have better alignment because this spot has no penumbra in the center-side quadrant.

Near the umbra in Ca II, there are pixels with high velocities (marked by an arrow), which are artifacts and occur due to failure of the method adopted for the Doppler-velocity determination. Similar Doppler-velocity and magnetic-field artifacts have been reported previously in measurements obtained from the *Michelson Doppler Imager* (Maurya and Ambastha, 2009; Maurya, 2010) and HMI (Maurya, Vemareddy, and Ambastha, 2012). We found that these pixels are associated with emission features of UFs marked by arrows in the intensity images. This problem was already described by Kneer, Mattig, and von Uexkuell (1981), and also by Tziotziou *et al.* (2006). UF emission features in Ca II can also be seen in Figure 5.

Figure 5 shows the intensity variations in Hα and Ca II lines as a function of time and wavelength for the three spatial locations in umbral, penumbral, and outer-penumbral regions. The Doppler velocities for bisector chords lengths of 0.35 Å (Hα) and 0.15 Å (Ca II) for corresponding positions are over-plotted (dotted curves). For position A, the Ca II line core shows that emission features of UFs and the Doppler velocities computed through bisectors may be inaccurate. Nevertheless, the Doppler velocities for all locations vary quasi-sinusoidally with time which represents the oscillatory motions of the chromospheric plasma emitting the Hα and Ca II lines. There is an interesting pattern of different periods corresponding to different locations. The period of oscillations increases from A to C.

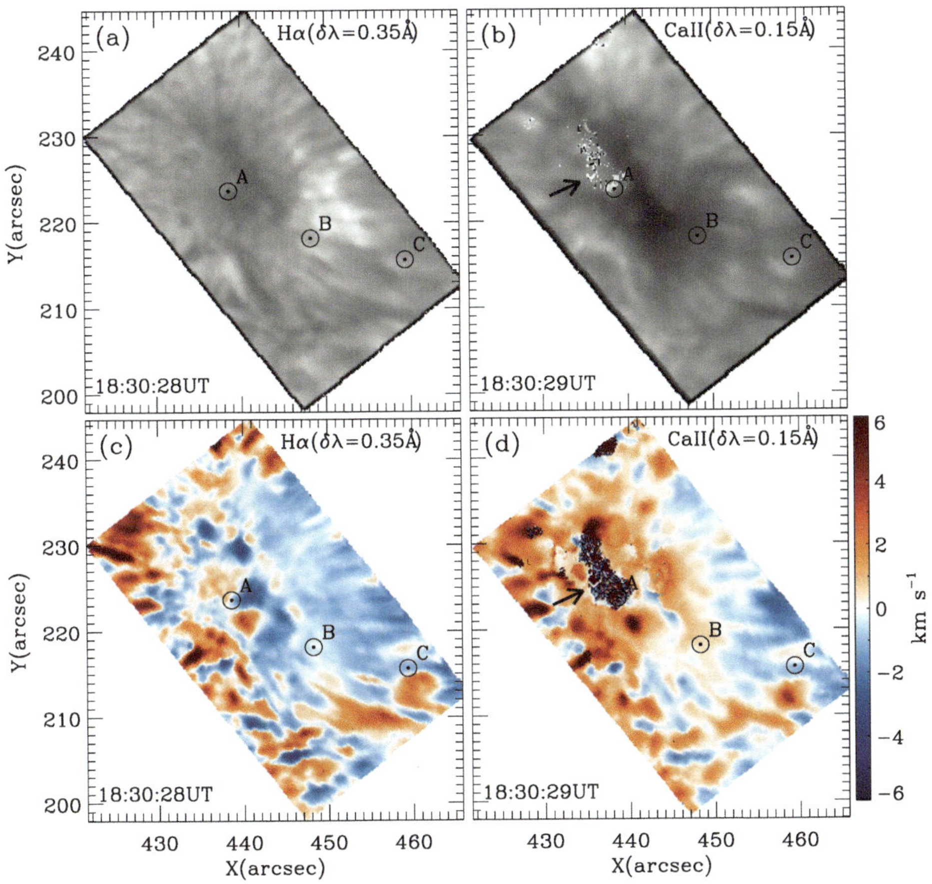

Figure 4 Simultaneous intensity (upper row) and Doppler-velocity (lower row) maps of AR NOAA 11242 computed using the bisector method for the Hα ($\delta\lambda = 0.35$ Å) and Ca II ($\delta\lambda = 0.15$ Å). Arrows mark the umbral locations where Doppler-velocity measurement in Ca II failed due to emission features in the umbral region. The positions A, B, and C, labeled with circles, represent the locations for which wavelength–time maps are shown in Figure 5.

Figure 6 shows space–time intensity maps (upper row) for Hα − 0.35 Å (a) and Ca II − 0.15 Å (b), and Doppler maps (lower row) for the Hα ($\delta\lambda = 0.35$ Å) (c) and Ca II ($\delta\lambda = 0.15$ Å) (d) along the line PQ (see Figure 1a). It is evident that the contrast of the intensity and Doppler-velocity variations in the umbral–penumbral region is large. Also the contrast in Hα is larger than in Ca II. In the umbral region, the ridges are almost vertical and show sinusoidal patterns of intensity and Doppler variations with time. Near the umbral boundary, there is a sharp change in the slope of the ridges. The intensity and Doppler-velocity contrasts in the ridges decay with distance from the umbral boundary to the outer edge of the penumbra, and it is difficult to distinguish intensity ridges further. This shows that the RPWs decay in the super-penumbral regions of the sunspot.

From the raster images, we found that the amplitude of RPWs gradually decreases away from the line centers. It is difficult to identify the RPWs beyond the 0.75 Å in the Hα and 0.50 Å in the Ca II. Also, the amplitude of RPWs decreases from the umbral boundary to the

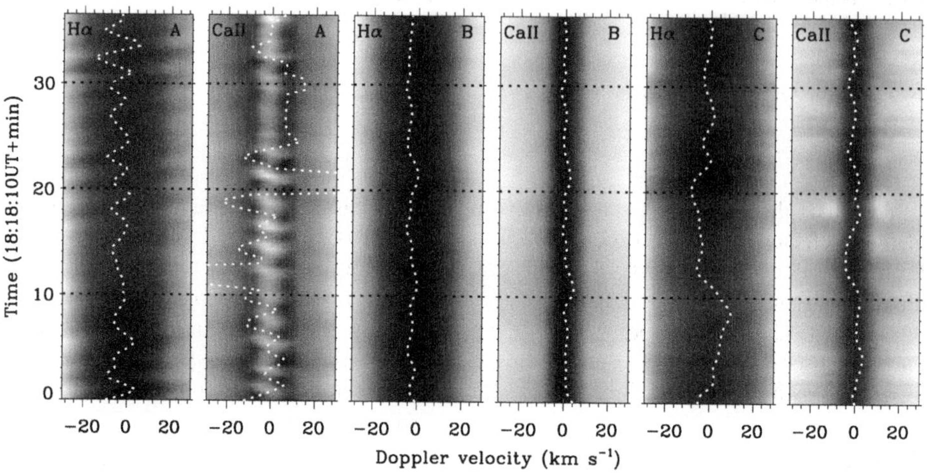

Figure 5 Intensity as a function of wavelength and time for Hα (panels, first, third, and fifth from the left) and Ca II (panels, second, fourth and sixth from the left) at three spatial locations [A, B, and C] respectively for the umbra, penumbra, and super-penumbra (see Figure 4). Dotted curves, passing through the lines' centers, show the Doppler velocities (magnified by factor of five for all of the panels except panel 2 from the left) at corresponding locations.

outer edge of the penumbra, which is consistent with the earlier reports (Giovanelli, 1972; Zirin and Stein, 1972).

The space–time Doppler panels of Figure 6 show that the RPWs have much more blue than red in Hα, the reverse in Ca II but no such asymmetry appears in Tziotziou *et al.* (2006, 2007). However, their traces also have larger amplitude for Ca II in the umbra and a shift between largely negative for Hα and largely positive for Ca II.

Figure 7 shows a sample of oscillation-power maps, for the Hα ($\delta\lambda = 0.35$ Å) (upper) and Ca II ($\delta\lambda = 0.15$ Å) (lower), averaged over different frequency bands. For comparison, all the maps are shown in the same amplitude range. It is evident that there is more power in the sunspot umbra at higher frequency band ($4.5 \le \nu < 7.5$ mHz) than the lower-frequency band ($1.5 \le \nu < 4.5$ mHz), while in the outer penumbra the power is higher at lower frequencies than at higher frequencies. This is consistent with earlier reports: *e.g.* Rouppe van der Voort *et al.* (2003). This pattern exists in both the spectral bands. The high-frequency oscillations in Ca II are found to be concentrated in smaller regions than in Hα, although there are some bad data points in the umbra of Ca II maps due to artifacts in the Doppler-velocity images (*cf.* Figure 4).

Figure 8 shows the oscillation power as a function of distance (along the line PQ in Figure 1a) and frequency. There is a clear linear trend in the maximum-power variation from the umbra to outward. The oscillations in the three- and five-minute bands, along with the features at other frequencies, can be seen in both maps.

In order to study the nature of the waves in the sunspot, we performed a Fourier phase difference analysis between the two Doppler signals obtained for Hα and Ca II. The Doppler-shift measurements along PQ were interpolated to a fixed-interval (30 second) sampling for this purpose.

Figure 9 shows the phase difference between the Hα ($\delta\lambda = 0.35$ Å) and Ca II ($\delta\lambda = 0.15$ Å) Doppler velocities as function of frequency for four locations in the umbra (a), penumbra (b), and super-penumbra (c). The average phase difference in the three- and five-millihertz frequency bands are around 40° where cross-spectral power is significant. The

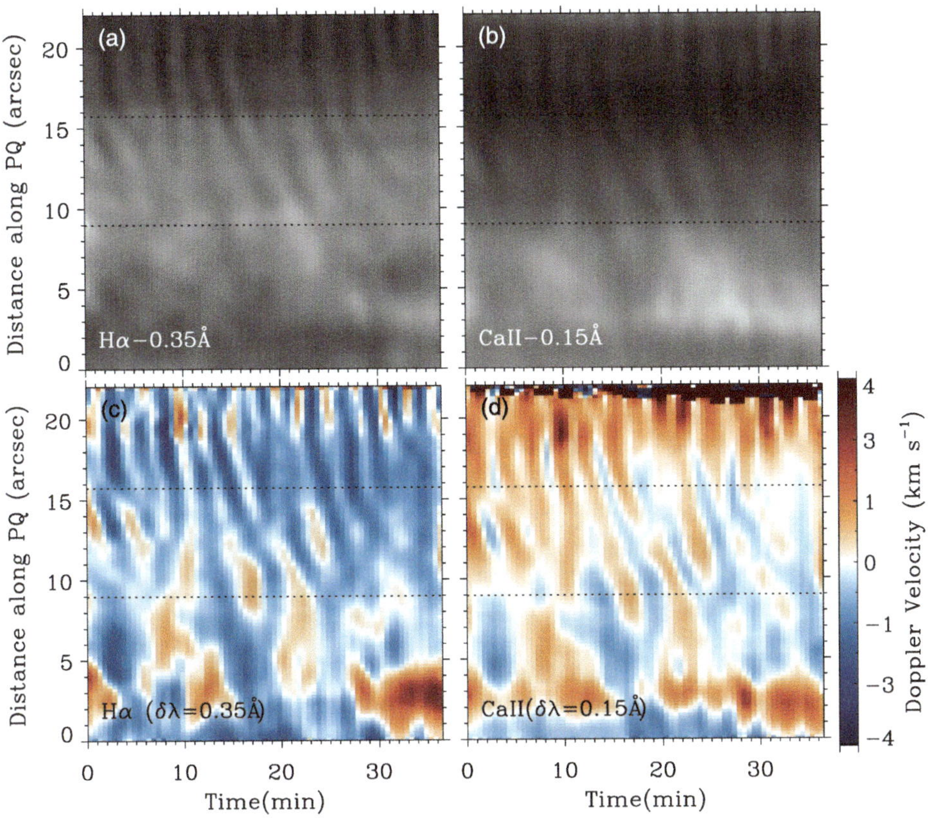

Figure 6 Space–time maps of intensity (upper row) and Doppler velocity (lower row), in Hα (left column) and Ca II (right column), along the reference line PQ (see Figure 1a). Dotted horizontal lines around 16″ and 9″ represent the approximate boundaries of the umbra and the penumbra, respectively.

phase difference shows that the UFs and RPWs are caused by upward-propagating MHD waves. The phase difference in the super-penumbral region is about 10° at lower frequency ≈ 1 mHz where significant power exists. This difference may be caused by noise and unequal sampling in the original data.

In order to ascertain the spatial variation in the oscillation power from umbra going outward, we have computed the root-mean-square (RMS) velocity [v_{rms}] along the line PQ; results are shown in Figure 10. It is evident that v_{rms} is large in the umbra of both the Hα ($\delta\lambda = 0.35$ Å) and Ca II ($\delta\lambda = 0.15$ Å) and decreases (increases) in the penumbra (super-penumbra). In the inner umbra, Ca II v_{rms} for point Q rises considerably above the 2.1 km s^{-1} axis cutoff, which is caused by the combined effects of high Doppler velocities and large uncertainty in the Doppler-shift measurements. In the outer umbra, v_{rms} of Hα is larger than v_{rms} of Ca II. This may be due to the difference in formation height between the two spectral lines or some other difference in response. For example, Ca II may have higher opacity in cool post-shock gas than Hα (which has high opacity only in hot gas), so that Ca II is sensitive to cool post-shock downdrafts and Hα is not. The K_{2V}-like shock-grain pattern in Ca II may arise from shock interference along the line of sight. Towards P the LOS alignment is lost. The large v_{rms} in the umbra consists mostly of high-frequency power. The high-frequency components likely describe sawtooth shocks (see Figure 5).

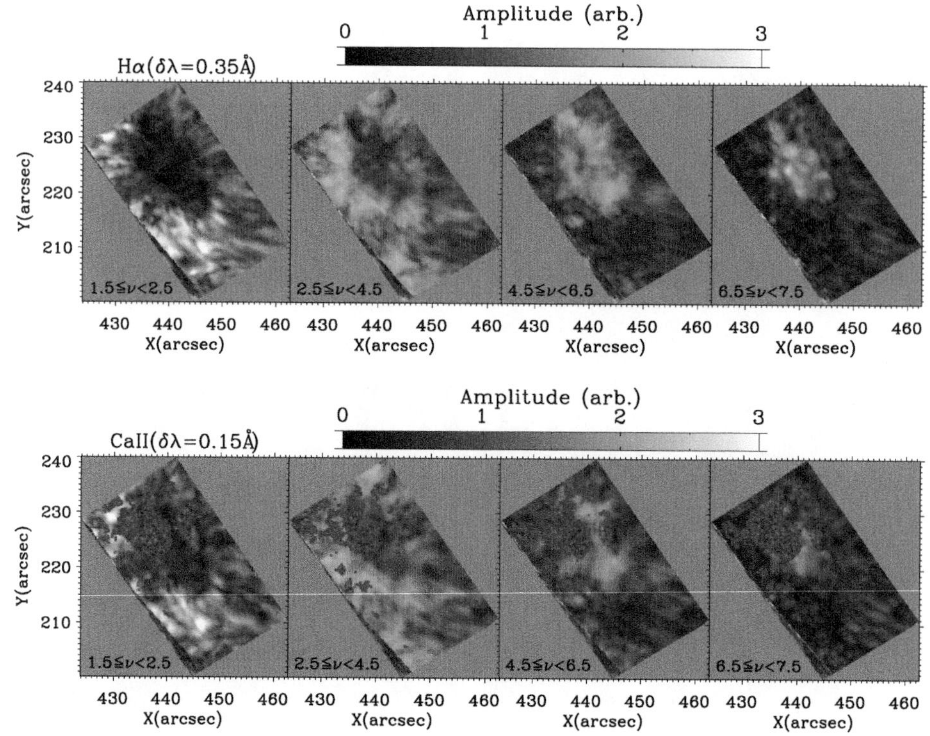

Figure 7 Average oscillation power maps in different frequency bands in mHz for the Hα ($\delta\lambda = 0.35$ Å) (upper) and Ca II ($\delta\lambda = 0.15$ Å) (lower).

Figure 11a shows the association in oscillation power and corresponding magnetic-field strength B from a rectangular area along line PQ of width of ten pixels. We noticed that the field strength of Figure 11a is not the same as in the HMI magnetogram in Figure 1. That indeed shows apparent polarity reversal in the penumbra from the line-of-sight projection. The mean oscillation power (see solid curves) in different frequency bands of both the Hα and Ca II lines shows a similar trend to the field strength [B]. At lower frequencies ($1.5 \leq \nu < 2.5$ mHz), the power is largest in areas with smaller B, which correspond to the regions of the outer penumbra, and rapidly decreases with B. In the frequency band $2.5 \leq \nu \leq 4.5$ mHz, the power initially increases up to $B \approx 700$ G and then decreases with B. In the frequency band $4.5 \leq \nu \leq 6.5$ mHz the oscillation power increases with B, becomes maximum around 1600 G, and then decreases rapidly. In the frequency band $6.5 \leq \nu \leq 7.5$ mHz most of the power is concentrated near the regions of very high B (> 1500 G), corresponding to the umbral region of the sunspot, while the power is nearly constant for the B range 300 – 1200 G.

To study the association between the oscillations and the field inclinations, we computed the inclination from the vertical to the surface using relation $\gamma = \cos^{-1}(B_r/|\mathbf{B}|)$, where B_r is the radial component of the magnetic field $\mathbf{B}$. The uniform-shear method is used (Moon *et al.*, 2003) to resolve the 180° ambiguity in the azimuth angle.

Figure 11b shows the variation in the oscillation power with magnetic-field inclination [γ] corresponding to the top panel of magnetic-field strength [B]. It is evident that the trend in power with inclination angle is the reverse of that with the field strength. The average

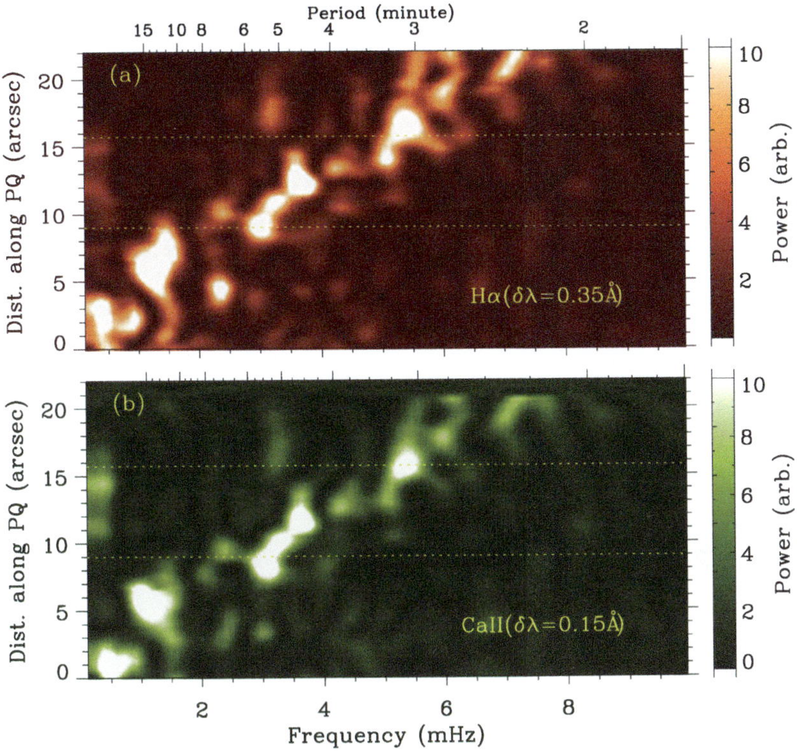

Figure 8 Oscillation-power maps along the line PQ (see Figure 1a) for the Doppler velocities of Hα ($\delta\lambda = 0.35$ Å) and Ca II ($\delta\lambda = 0.15$ Å). Dotted lines mark the boundaries as shown in Figure 6.

oscillation power (solid curves) in different frequency bands of both the Hα and Ca II lines behave similarly to the inclination except for smaller $\gamma \lessapprox 40$ at the high-frequency band $6.5 \leq \nu \leq 7.5$ mHz. This discrepancy in the Hα and Ca II lines arises due to UFs in Ca II. In the lower-frequency band ($1.5 \leq \nu < 2.5$ mHz), most power is concentrated at higher $\gamma > 80°$. The peak of the power distribution shifts toward lower inclination for higher-frequency bands. For instance in the frequency band $6.5 \leq \nu < 7.5$ mHz, most of the power is concentrated in the inclination region $10° - 30°$.

The opposite relation of inclination and field strength to the power is obvious from the magnetic-field distribution in the sunspot umbra; we have a strong field with small inclination. From the umbra outward the field strength increases while inclination decreases. Figure 11 shows that the oscillation power is associated both with the field strength and the inclination angle, but they behave differently in different frequency bands.

5. Summary and Conclusions

We studied the nature of running penumbral waves using Hα and Ca II Doppler images constructed with a bisector method from area scans with the FISS spectrometer. We found that the RPWs are easily seen in the intensity images constructed near the core, and their amplitude decays with wavelength from the line center outward. Also, their amplitude decreases from the umbra outward. These results are consistent with earlier results

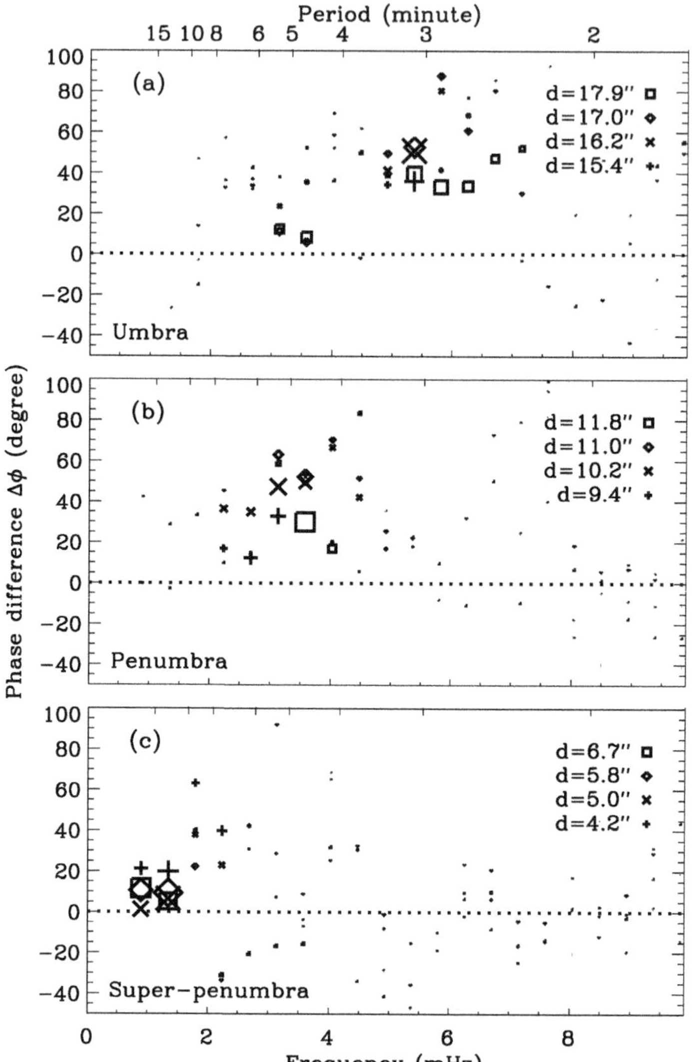

Figure 9 Phase difference [$\Delta\phi$] in Ca II ($\delta\lambda = 0.15$ Å) and Hα ($\delta\lambda = 0.35$ Å) Doppler velocities of: (a) umbra, (b) penumbra, and (c) super-penumbra, at four locations marked by symbols, square (□), diamond (◇), cross (×), and plus (+) along the reference line PQ (Figure 10a), where Ca II ($\delta\lambda = 0.15$ Å) is taken as leading. The distance [d] is measured from the point P along PQ. The symbol sizes represent the cross-spectral power between the two Doppler time series.

about the RPWs (Zirin and Stein, 1972; Christopoulou, Georgakilas, and Koutchmy, 2000; Tziotziou *et al.*, 2006). The running penumbral waves decay near the outer boundary of the penumbra.

We found that the chromosphere umbra of the sunspot shows large RMS velocity [v_{rms}] in both spectral lines. From the umbra going outward, the v_{rms} gradually decreases, and there is no distinct boundary between the v_{rms} of the umbra and penumbra; however, v_{rms} has a minimum in the penumbral region. There is another interesting pattern seen in the v_{rms} of

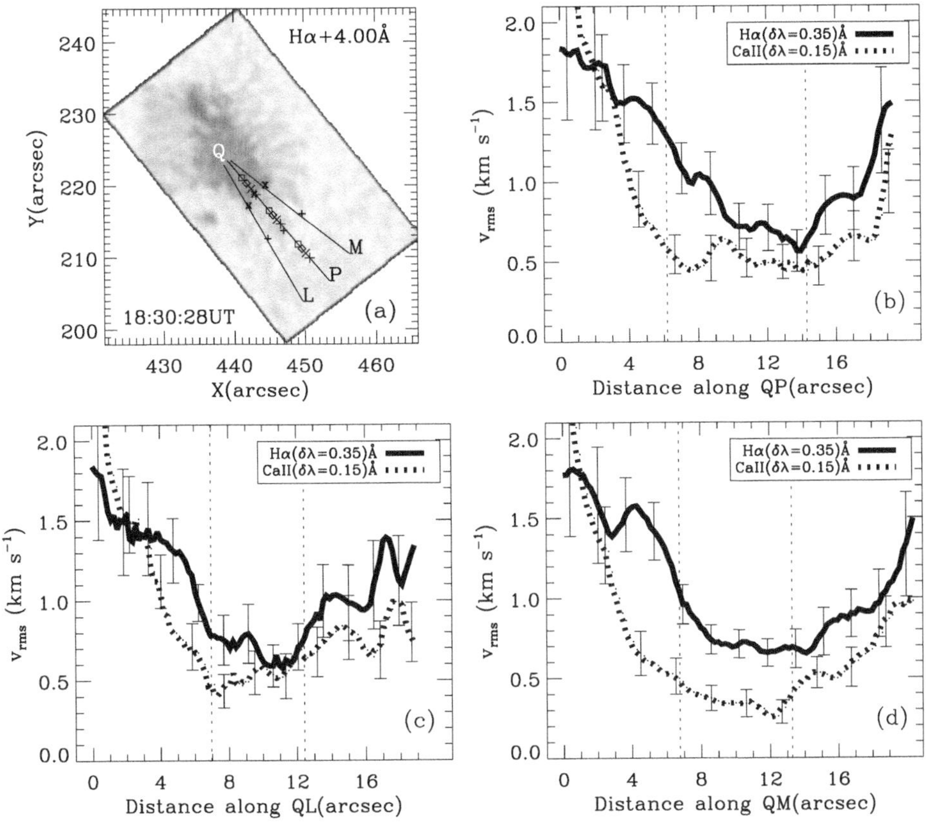

Figure 10 Variation in v_{rms} velocities of Hα ($\delta\lambda = 0.35$ Å) and Ca II ($\delta\lambda = 0.15$ Å) along the reference lines QN(b), QM(d), and QL(c) marked in the panel (a). Vertical dotted lines show the approximate umbral and penumbral boundaries marked by cross (×) and plus (+) symbols, respectively, over the solid curves in the panel (a). Other symbols (square–diamond–cross–plus "□ ◇ ×+") in the umbra, penumbra, and super-penumbra of the sunspot mark the locations along the line PQ (same as in Figure 1a) for phase analysis in Figure 9.

Hα and Ca II lines; v_{rms} is smaller in Ca II than in Hα away from the umbral center. These results reveal that the Hα and Ca II lines are formed differently in the shocks that make up UFs. Furthermore v_{rms} is related to the oscillation power [power $\propto v_{\mathrm{rms}}^2$]. That is, we have larger power in the umbra than in penumbra; also, the power is larger in Hα than in Ca II.

Our time-series analysis of chromospheric Doppler maps shows high-frequency power in the umbra of both the spectral lines, Hα and Ca II, which is also evident from the v_{rms} distribution. This study shows that the total power gradually decreases with frequency from the umbra outward which confirms earlier reports (Tziotziou *et al.*, 2006; Nagashima *et al.*, 2007; Socas-Navarro *et al.*, 2009). The power maps indicate that this decrease is set by the high-frequency contribution. Our analysis of spectral observations shows that the high-frequency oscillations exist in the umbra of five-minute band. The observed power in the penumbral region shows strong peak in the three-minute band while the power is small at other frequencies.

The penumbral oscillations are regarded as a tail of the five-minute oscillations resulting in further dependence of the acoustic cutoff frequency on the magnetic-field inclination

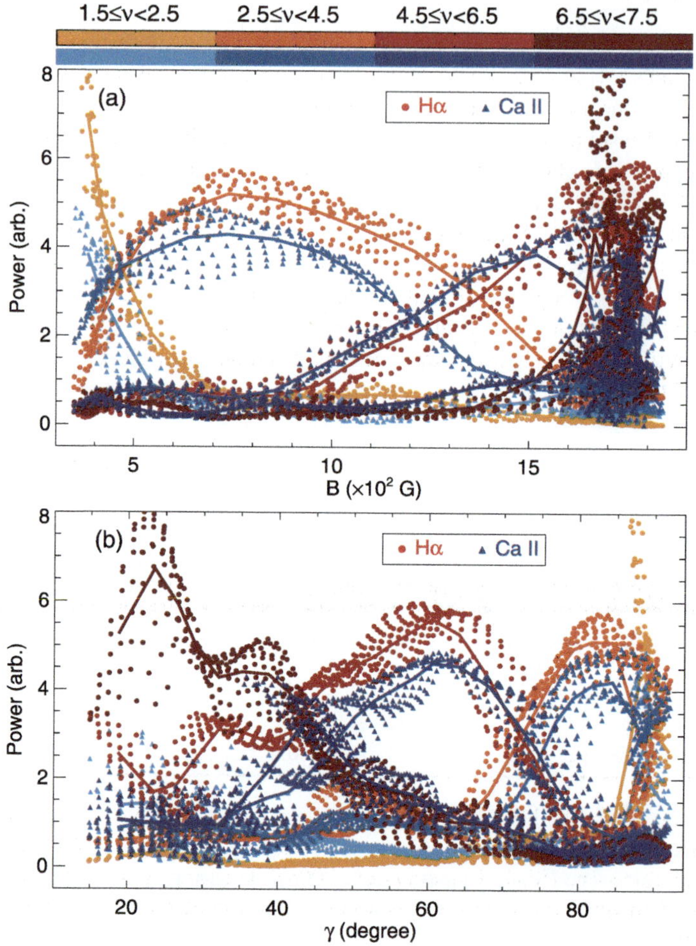

Figure 11 Scatter plots of oscillation power in different frequency bands (shown with different color gradients) against (a) magnetic field strength [B], and (b) magnetic-field inclination [γ], with mean curves (solid lines), for the rectangular area along line PQ of width ten pixels (see Figure 1a). Red and blue colors correspond to the Hα and Ca II lines, respectively.

(Cally, Bogdan, and Zweibel, 1994; De Pontieu, Erdélyi, and James, 2004). The oscillation power shows a strong relationship with the magnetic-field strength and angle of inclination. Our analysis showed that the peak oscillation frequency depends not only upon the inclination but also on the field strength.

Observations in Hα and Ca II bands demonstrated that the oscillations with frequencies around 5.0 mHz are dominant in the umbral and inner-penumbral regions, together with the presence of RPWs in the three-millihertz band.

We can summarize the above results as follows:

- Ca II showed reverse C-shaped bisectors at some locations within the sunspot.
- Bisector-measured Doppler velocity in Ca II is affected by umbral emission features.
- The amplitude of RPWs decreases with distance from the umbra outward and decays near the penumbral boundary.

- RPWs have much more blue than red in Hα and the reverse in Ca II.
- Peak-power frequency gradually increases from umbra to outward in both lines.
- In umbra and penumbra of the sunspot, the Hα and Ca II lines showed a phase difference of $\approx 40^\circ$ in the three- and five-minute oscillation bands.
- v_{rms} in the umbra is larger than in the penumbra of both lines. In the outer umbra, v_{rms} is larger in Hα than v_{rms} in Ca II.
- The oscillation power shows different relations with the magnetic-field strength and the inclination in different frequency bands.

We conclude that the main oscillation properties still suggest MHD waves propagating upward along fanning field, as in earlier studies, but with interesting as yet unexplained response differences between the two lines. The transformation of vertical oscillations into RPWs, at the boundary of umbra and penumbra, is an interesting feature, which requires further observations and theoretical studies to explain the wave phenomenon in the sunspot.

Acknowledgements HMI data are courtesy of NASA/SDO and the HMI science team. We thank the anonymous referee for their comments and important suggestions that helped to improve the quality of the manuscript. This work was support by the National Research Foundation of Korea (2011-0028102).

References

Bard, S., Carlsson, M.: 2010, *Astrophys. J.* **722**, 888. ADS:2010ApJ...722..888B. doi:10.1088/0004-637X/722/1/888.

Beckers, J.M., Tallant, P.E.: 1969, *Solar Phys.* **7**, 351. ADS:1969SoPh....7..351B. doi:10.1007/BF00146140.

Bhatnagar, A., Livingston, W.C., Harvey, J.W.: 1972, *Solar Phys.* **27**, 80. ADS:1972SoPh...27...80B. doi:10.1007/BF00151772.

Bloomfield, D.S., Lagg, A., Solanki, S.K.: 2007, *Astrophys. J.* **671**, 1005. ADS:2007ApJ...671.1005B. doi:10.1086/523266.

Cally, P.S., Bogdan, T.J., Zweibel, E.G.: 1994, *Astrophys. J.* **437**, 505. ADS:1994ApJ...437..505C. doi:10.1086/175014.

Cavallini, F., Ceppatelli, G., Righini, A., Alamanni, N.: 1986, *Astron. Astrophys.* **156**, 310. ADS:1986A&A...156..310C.

Chae, J., Park, H.M., Ahn, K., Yang, H., Park, Y.D., Nah, J., Jang, B.H., Cho, K.S., Cao, W., Goode, P.R.: 2012, *Solar Phys.* ADS:2012SoPh..tmp..248C. doi:10.1007/s11207-012-0147-x.

Christopoulou, E.B., Georgakilas, A.A., Koutchmy, S.: 2000, *Astron. Astrophys.* **354**, 305. ADS:2000A&A...354..305C.

De Pontieu, B., Erdélyi, R., James, S.P.: 2004, *Nature* **430**, 536. ADS:2004Natur.430..536D. doi:10.1038/nature02749.

Dravins, D., Lindegren, L., Nordlund, A.: 1981, *Astron. Astrophys.* **96**, 345. ADS:1981A&A....96..345D.

Giovanelli, R.G.: 1972, *Solar Phys.* **27**, 71. ADS:1972SoPh...27...71G. doi:10.1007/BF00151771.

Gray, D.F.: 1976, *The Observation and Analysis of Stellar Photospheres*, Wiley-Interscience, New York. ADS:1976oasp.book.....G.

Havnes, O.: 1970, *Solar Phys.* **13**, 323. ADS:1970SoPh...13..323H. doi:10.1007/BF00153554.

Keil, S.L., Yackovich, F.H.: 1981, *Solar Phys.* **69**, 213. ADS:1981SoPh...69..213K. doi:10.1007/BF00149989.

Kneer, F., Mattig, W., von Uexkuell, M.: 1981, *Astron. Astrophys.* **102**, 147. ADS:1981A&A...102..147K.

Lomb, N.R.: 1976, *Astrophys. Space Sci.* **39**, 447. ADS:1976Ap%26SS..39..447L. doi:10.1007/BF00648343.

Maurya, R.A.: 2010, PhD thesis, Udaipur Solar Observatory/Physical Research Laboratory, Mohan Lal Sukhadiya Univ., Udaipur, India. ADS:2010PhDT.........2M.

Maurya, R.A., Ambastha, A.: 2009, *Solar Phys.* **258**, 31. ADS:2009SoPh..258...31M. doi:10.1007/s11207-009-9397-7.

Maurya, R.A., Vemareddy, P., Ambastha, A.: 2012, *Astrophys. J.* **747**, 134. ADS:2012ApJ...747..134M. doi:10.1088/0004-637X/747/2/134.

Moon, Y., Wang, H., Spirock, T.J., Goode, P.R., Park, Y.D.: 2003, *Solar Phys.* **217**, 79. ADS:2003SoPh..217...79M.

Moore, R.L.: 1981, *Space Sci. Rev.* **28**, 387. ADS:1981SSRv...28..387M. doi:10.1007/BF00212601.

Nagashima, K., Sekii, T., Kosovichev, A.G., Shibahashi, H., Tsuneta, S., Ichimoto, K., Katsukawa, Y., Lites, B., Nagata, S., Shimizu, T., Shine, R.A., Suematsu, Y., Tarbell, T.D., Title, A.M.: 2007, *Publ. Astron. Soc. Japan* **59**, 631. ADS:2007PASJ...59S.631N.
Nye, A.H., Thomas, J.H.: 1974, *Solar Phys.* **38**, 399. ADS:1974SoPh...38..399N. doi:10.1007/BF00155077.
Nye, A.H., Thomas, J.H.: 1976, *Astrophys. J.* **204**, 582. ADS:1976ApJ...204..582N. doi:10.1086/154206.
Pearson, K.: 1901, *Phil. Mag.* **2**, 559.
Rees, D.E., López Ariste, A., Thatcher, J., Semel, M.: 2000, *Astron. Astrophys.* **355**, 759. ADS:2000A&A...355..759R.
Rouppe van der Voort, L.H.M., Rutten, R.J., Sütterlin, P., Sloover, P.J., Krijger, J.M.: 2003, *Astron. Astrophys.* **403**, 277. ADS:2003A&A...403..277R. doi:10.1051/0004-6361:20030237.
Scargle, J.D.: 1982, *Astrophys. J.* **263**, 835. ADS:1982ApJ...263..835S. doi:10.1086/160554.
Schou, J., Scherrer, P.H., Bush, R.I., Wachter, R., Couvidat, S., Rabello-Soares, M.C., Bogart, R.S., Hoeksema, J.T., Liu, Y., Duvall, T.L., Akin, D.J., Allard, B.A., Miles, J.W., Rairden, R., Shine, R.A., Tarbell, T.D., Title, A.M., Wolfson, C.J., Elmore, D.F., Norton, A.A., Tomczyk, S.: 2012, *Solar Phys.* **275**, 229. ADS:2012SoPh..275..229S. doi:10.1007/s11207-011-9842-2.
Slaughter, C.D., Wilson, A.M.: 1972, *Solar Phys.* **24**, 43. ADS:1972SoPh...24...43S. doi:10.1007/BF00231081.
Socas-Navarro, H., McIntosh, S.W., Centeno, R., de Wijn, A.G., Lites, B.W.: 2009, *Astrophys. J.* **696**, 1683. ADS:2009ApJ...696.1683S. doi:10.1088/0004-637X/696/2/1683.
Stebbins, R., Goode, P.R.: 1987, *Solar Phys.* **110**, 237. ADS:1987SoPh..110..237S. doi:10.1007/BF00206421.
Tziotziou, K., Tsiropoula, G., Mein, P.: 2002, *Astron. Astrophys.* **381**, 279. ADS:2002A&A...381..279T. doi:10.1051/0004-6361:20011419.
Tziotziou, K., Tsiropoula, G., Mein, N., Mein, P.: 2006, *Astron. Astrophys.* **456**, 689. ADS:2006A&A...456..689T. doi:10.1051/0004-6361:20064997.
Tziotziou, K., Tsiropoula, G., Mein, N., Mein, P.: 2007, *Astron. Astrophys.* **463**, 1153. ADS:2007A&A...463.1153T. doi:10.1051/0004-6361:20066412.
Uitenbroek, H.: 2006, *Astrophys. J.* **639**, 516. ADS:2006ApJ...639..516U. doi:10.1086/499220.
von Uexküll, M., Kneer, F., Mattig, W.: 1983, *Astron. Astrophys.* **123**, 263. ADS:1983A&A...123..263V.
Zhugzhda, I.D., Dzhalilov, N.S.: 1984, *Astron. Astrophys.* **133**, 333. ADS:1984A&A...133..333Z.
Zirin, H., Stein, A.: 1972, *Astrophys. J. Lett.* **178**, L85. ADS:1972ApJ...178L..85Z. doi:10.1086/181089.

DOI 10.1007/978-3-319-12123-9_6
Reprinted from *Solar Physics* Journal, DOI 10.1007/s11207-013-0313-9

Doppler Shifts of the Hα Line and the Ca II 854.2 nm Line in a Quiet Region of the Sun Observed with the FISS/NST

Jongchul Chae · Hyung-Min Park · Kwangsu Ahn · Heesu Yang · Young-Deuk Park · Kyung-Suk Cho · Wenda Cao

Received: 13 August 2012 / Accepted: 24 April 2013 / Published online: 5 June 2013
© Springer Science+Business Media Dordrecht 2013

Abstract The characteristics of Doppler shifts in a quiet region of the Sun are compared between the Hα line and the Ca II infrared line at 854.2 nm. A small area of $16'' \times 40''$ was observed for about half an hour with the Fast Imaging Solar Spectrograph (FISS) of the 1.6 meter New Solar Telescope (NST) at Big Bear Solar Observatory. The observed area contains a network region and an internetwork region, and identified in the network region are fibrils and bright points. We infer Doppler velocity v_{m} from each line profile at each individual point with the lambdameter method as a function of half wavelength separation $\Delta\lambda$. It is confirmed that the bisector of the spatially averaged Ca II line profile has an inverse C-shape with a significant peak redshift of $+1.8\ \mathrm{km\,s^{-1}}$. In contrast, the bisector of the spatially averaged Hα line profile has a C-shape with a small peak blueshift of $-0.5\ \mathrm{km\,s^{-1}}$. In both lines, the bisectors of bright network points are significantly redshifted not only at the line centers, but also at the wings. The Ca II Doppler shifts are found to be correlated with the Hα ones with the strongest correlation occurring in the internetwork region. Moreover, we find that here the Doppler shifts in the two lines are essentially in phase. We discuss the physical implications of our results in view of the formation of the Hα line and Ca II 854.2 nm line in the quiet region chromosphere.

Keywords Chromosphere, quiet · Spectral line, intensity and diagnostics

Initial Results from FISS
Guest Editor: Jongchul Chae

J. Chae (✉) · H.-M. Park · H. Yang
Astronomy Program, Department of Physics and Astronomy, Seoul National University, Seoul 151-747, Korea
e-mail: chae@astro.snu.ac.kr

Y.-D. Park · K.-S. Cho
Korea Astronomy and Space Science Institute, Daejeon, Korea

K. Ahn · W. Cao
Big Bear Solar Observatory, New Jersey Institute of Technology, 40386 North Shore Lane, Big Bear City, CA 92314, USA

1. Introduction

The chromosphere – traditionally referring to the layer seen outside the solar limb that emits intense light in the Hα line and other very strong lines in the visible and near-infrared spectra – still remains mysterious in many aspects (Rutten, 2010). This layer, particularly its outer part, comprises a number of inhomogeneous plasma features relevant to flows, waves, shocks and magnetic fields.

The simultaneous recording of the Hα line and one of the Ca II lines is a useful diagnostic of these features, particularly if their Doppler absorption widths can be measured. Because of the big difference in atomic mass, the absorption width of the Hα line is dominated by thermal broadening, and that of the Ca II line, by non-thermal broadening, which allows us to separately determine temperature and the speed of non-thermal motion. This kind of atomic weight tool was previously exploited by Scoas-Navarro and Elmore (2005) with the Ca II infrared lines and the He I line at 1083 nm, and by Cauzzi *et al.* (2009) with the Hα line and Ca II 854.2 nm line.

It is quite worthwhile to investigate the characteristics of Doppler shift in the chromosphere by comparing between the Hα line and the Ca II line. We expect that the Doppler shifts in the two lines would be the same or at least be correlated with each other, if the two lines are absorbed by the same volume in the chromosphere, a prerequisite for the atomic weight tool above to be applicable. As a matter of fact, comparing Doppler shifts between different chromospheric lines is not new in solar physics (*e.g.* Cram, Brown, and Beckers, 1977), but is recently becoming important again with the arrival of new instruments like IBIS (*e.g.* Cauzzi *et al.*, 2009) or CRISP (*e.g.* Sekse, Rouppe van der Voort, and De Pontieu, 2012).

Different methods have been used to infer Doppler shifts from the chromospheric lines. The early measurements were done with the Dopplergram technique that compares intensity between two wavelengths $\pm \Delta\lambda$ offset from the center of the reference line as in the case of weaker lines of photospheric origin (Leighton, Noyes, and Simon, 1962). Since the full line profile became available, the wavelength of minimum intensity or line center has frequently been used as a measure of Doppler shift (Cram, Brown, and Beckers, 1977; Lites, Rutten, and Kalkofen, 1993; Cauzzi *et al.*, 2008). More general, but less widely used is the so-called lambdameter method (Deubner, 1974; Mein and Mein, 1976; Deubner, Waldschik, and Steffens, 1996). For given length $2\Delta\lambda$, this method seeks the wavelength of the middle point of a horizonal chord of the given length that fits into the line profile. It usually uses the inner wings to infer Doppler shift instead of the line center. This method yields the intensity of the chord as well, and the series of such wavelength-intensity pairs obtained for different values of $\Delta\lambda$ naturally leads to the construction of the bisector of the line profile.

The bisector of a line profile provides much information on the velocity field and the formation of the line. It has been well known that the bisector of a strong photospheric line is usually shaped like a distorted C, which is successfully explained by the combined effect of hot rising material and cool falling material in the granulation together with the fact that the core of a strong line is formed high in the photosphere above the granulation (Gray, 1992; Asplund *et al.*, 2000). The bisectors of chromospheric lines, however, are not well established observationally, and even little understood. It is only several years ago that Uitenbroek (2006) first reported an interesting finding that "spatially averaged intensity profiles of the chromospheric Na I D and Ca II infrared lines exhibit a pronounced red asymmetry in their cores with bisectors in the shape of an inverse C." This finding is contrasted with the well-known C-shape in the bisectors of spectral lines of photospheric origin.

In this paper we investigate the observed characteristics of Doppler shifts of the Hα line and Ca II 854.2 nm line in different kinds of features in a quiet region of the Sun. Using the

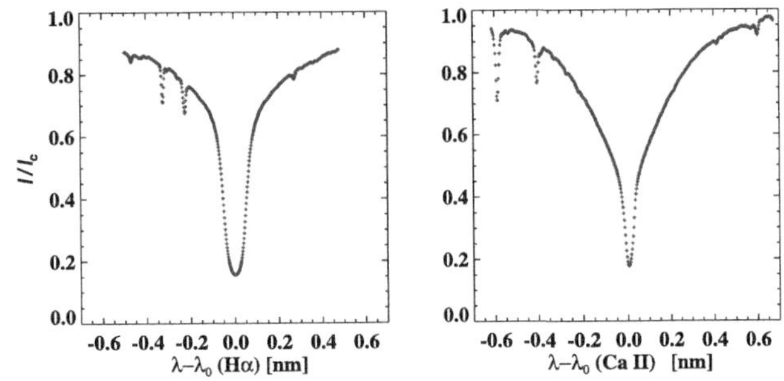

Figure 1 Quiet Sun profiles of the Hα (left) and Ca II 854.2 nm lines (right) obtained by averaging over 16″ × 40″ and over about half an hour. Telluric lines in the line cores were removed.

Fast Imaging Solar Spectrograph (FISS) of the 1.6 meter New Solar Telescope (NST) at Big Bear Solar Observatory, we observed a quiet region near the center of the solar disk for about half an hour. The FISS is a dual band Echélle spectrograph that can record high-resolution spectra of the Hα line and the Ca II 854.2 nm on two different cameras simultaneously (Chae *et al.*, 2012).

We study observational properties of Doppler shifts that may serve as a basis for our future goal of determining the physical parameters of different kinds of features from the observed line profiles. The categories of the features we deal with are

i) all the features in the whole quiet region,
ii) all the features in the internetwork region,
iii) Hα fibrils and Ca II fibrils in the network region, and
iv) bright points in the network region.

For each group of the features, we examine characteristics of average Doppler shifts from the bisectors of spatially averaged spectral profile of each line, and the characteristics of their spatio-temporal fluctuations based on the RMS values and cross-correlations of Doppler shifts.

2. Observation and Analysis

2.1. Observation

The FISS is an imaging spectrograph that successively takes a pair of spectrograms at the dual bands at each location of the slit inside the field of view. With this instrument we observed a small quiet region of 16″ × 40″ on 13 July 2011. Each Hα line spectrogram comprises 512 × 256 pixels of 1.9 pm (picometer) × 0.16″, covering 0.97 nm (nanometer) × 40″, and each Ca II 854.2 nm line spectrogram, 502 × 250 pixels of 2.6 pm × 0.16″, covering 1.3 nm × 40″. Figure 1 presents the profiles of the two lines averaged over the observed quiet region.These are representative of the quiet Sun in that they are averaged over a disk-center area that includes both network and internetwork, and over a time span that is much longer than the predominant periods of solar oscillations.

The FISS does imaging based on the fast scan of the slit over the field of view. The step size was set to 0.16″, and each scan of 100 steps was done in 18 s. The scan was repeated for about half an hour. The resulting field of view of the observed region is 16″ × 40″ or 12 Mm × 29 Mm. Figure 2 shows the raster images constructed from a set of one scan data at the fixed wavelengths. The figure shows that even though small, the field of view is still big enough to contain typical network features as well as internetwork features. The

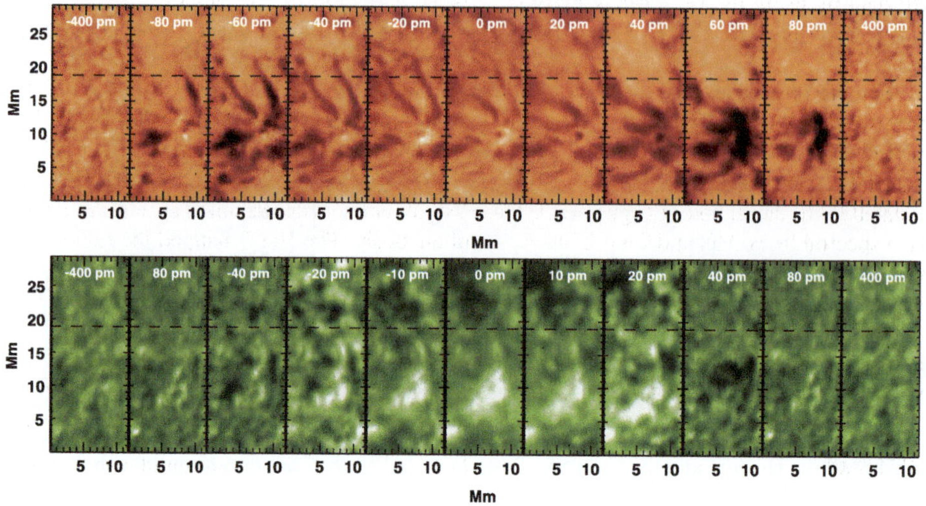

Figure 2 Images of the observed region constructed from one raster scan at different wavelengths of the Hα line (upper) and the Ca II 854.2 nm line (lower). The horizontal dashed lines indicate the boundary between the network region and the internetwork region.

early run performance of the FISS as well as basic data processing was described by Chae *et al.* (2012). The basic processing of data includes dark subtraction, flat fielding, distortion correction, and noise suppression.

The spatial resolution of the Hα data in this observation is estimated to be better than 1″, and that of the Ca II 854.2 nm data, to be poorer, as was reported by Chae *et al.* (2012). These resolutions fall short of the diffraction-limited resolutions of the NST – 0.10″ and 0.13″ – mainly for two reasons. First, the troublesome seeing is to be corrected by the adaptive optics (AO) only with post image processing being impossible, but the currently operating AO is not optimized for visible wavelengths and quiet region observations. In addition, we realize that the relay optics of the FISS in early observations that produced the current data was not optimal, being subject to ill-focusing and chromatic abberation. Upon the improvement of both the AO and the relay optics, future FISS observations will have much higher spatial resolution, even though not as high as diffraction-limited ones.

2.2. Wavelength Calibration

We have calibrated wavelengths by setting the rest frame to the center of the Sun. The adopted laboratory wavelengths of the Hα line and the Ca II line are 656.2817 nm and 854.2089 nm, respectively. In each of the Hα band and Ca II 854.2 nm band, this wavelength calibration is done in two steps. First, wavelength is calibrated in the rest frame of the Earth using two telluric lines in each band – the H_2O lines at 656.1097 nm and 656.4206 nm in the Hα band, and the H_2O lines at 854.0817 nm and 854.6222 nm in the Ca II band. Next, the radial velocity of the Sun relative to the Earth is calculated based on the ephemeris data of the Sun, and the corresponding Doppler shift is subtracted from the above wavelength values, resulting in the wavelengths calibrated in the rest frame tied to the center of the Sun. Our wavelength calibration is distinct from the conventional approach defining the zero of Doppler shift as the spatial average over the whole data set (Deubner and Fleck, 1990; Cauzzi *et al.*, 2008).

2.3. Removal of Telluric Lines

After wavelength calibration, we removed telluric lines near the line cores in the following steps. First, we identify the center of each telluric line, and define two small spectral regions surrounding the line. Second, the intensity variation in the two spectral regions is fitted by a regularly varying low-order polynomial function. Finally, the intensity variation in the local spectral region containing the telluric line is replaced by this function.

2.4. Lambdameter Method

With the observed spectral profile $I(\lambda)$ and half the length of the chord $\Delta\lambda$ given, the lambdameter method (Deubner, Waldschik, and Steffens, 1996) calculates the Doppler (line-of-sight) velocity v_{m} from the middle wavelength of the chord, λ_{m} that satisfies $I(\lambda_{\mathrm{m}} - \Delta\lambda) = I(\lambda_{\mathrm{m}} + \Delta\lambda) = I_{\mathrm{l}}$ where the intensity of the chord I_{l} is obtained as a byproduct. In this lambdameter method both I_{l} and λ_{m} are considered as functions of $\Delta\lambda$. The bisector of a spectral line is nothing but the plot of all the pairs of $I_{\mathrm{l}}(\Delta\lambda)$ *vs.* $\lambda_{\mathrm{m}}(\Delta\lambda)$ obtained with different values of $\Delta\lambda$. Note that $I_{\mathrm{l}}(\Delta\lambda)$ and $\lambda_{\mathrm{m}}(\Delta\lambda)$ are independent of each other.

3. Feature Identification

The comparison of Figures 2 and 3 shows that the two lines differ in prominently visible features. In the network region, fibrils are conspicuous in the Hα line, and bright network points, in the Ca II line. As is well known (*e.g.* Cauzzi *et al.*, 2009), the Hα line core intensity mostly reflects light-scattering features, while the Ca II line core intensity is sensitive to temperature. Specifically, Leenaarts, Carlsson, and Rouppe van der Voort (2012) found from their state-of-art modeling of the Hα line formation that the core intensity is correlated with the average formation height. The internetwork region (hereafter IN) contains roundish dark features seen either at Ca II + 20 pm or at − 20 pm, but not at both wavelengths. They hence represent the shift or distortion of the Ca II line profile due to the presence of material moving either upward or downward, for example, during the passage of shock waves (Vecchio, Cauzzi, and Reardon, 2009).

Can we identify bright network points in Hα and fibrils in Ca II in the network region? Leenaarts *et al.* (2006a) and Leenaarts *et al.* (2006b) showed that Hα blue wing images are the best proxy for photospheric magnetic features. Bright points, however, are not obvious in our Hα images, probably because overlying fibrils are blocking out bright network points below. Meanwhile, Cauzzi *et al.* (2008) discovered a number of fibrils in the Ca II line using high-resolution observations with IBIS. In our Ca II observations, however, fibrils are not clearly visible. This may be partly because the spatial resolution of our Ca II data is not high enough, and partly because the Ca II fibrils may be intrinsically short (Rouppe van der Voort *et al.*, 2009; Cauzzi *et al.*, 2009). Note that this network region is not an enhanced one hosting many stable long fibrils. Nevertheless, a careful look at Figures 2 and 3 near the Hα fibrils reveals a few fibril-like structures in our Ca II images particularly at the wavelengths of ± 20 pm. Like the Hα fibrils, these "Ca II fibrils" appear dark and somewhat elongated, and extend radially out from the bright points, even though not so clearly defined as the Hα fibrils are.

Our feature recognition was done in every set of images constructed from each raster scan, based on the ratio of $I_{\mathrm{l}}(\Delta\lambda)$ to the mean value of the quiet region. The choice of $\Delta\lambda$

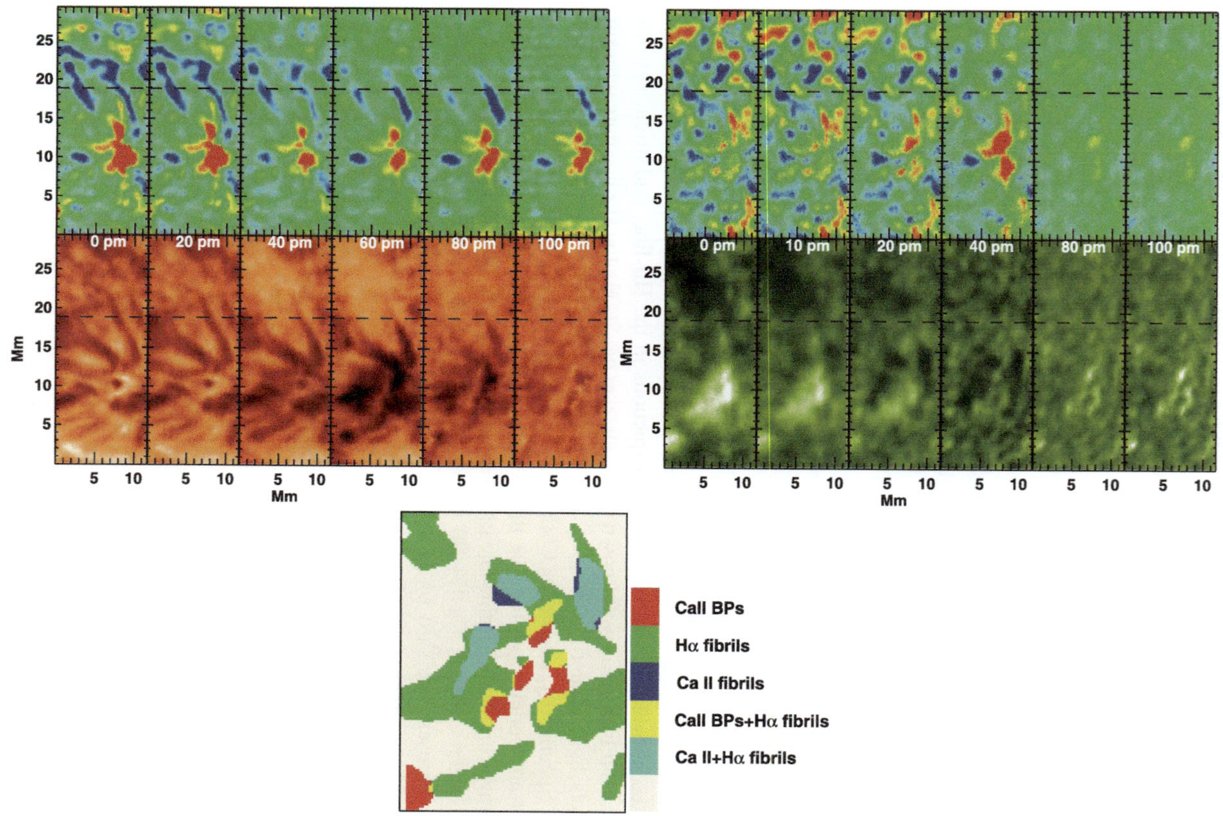

Figure 3 Top: Maps of v_{m} for different values of $\Delta\lambda$ from the Hα line (left) and the Ca II 854.2 nm line (right). For display the maps of v_{m} were scaled by $\pm 5\ \mathrm{km\,s^{-1}}$ (Hα) and $\pm 2.5\ \mathrm{km\,s^{-1}}$ (Ca II), respectively. Middle: Maps of I_1 constructed at the same wavelengths. Bottom: feature identification in the network region (below the horizontal line). Note that Ca II BPs identified here are well visible in the I_1 map of Ca II with $\Delta\lambda = 100$ pm, Hα fibrils, in the Hα map with $\Delta\lambda = 60$ pm, Ca II fibrils in the Ca II map with $\Delta\lambda = 20$ pm.

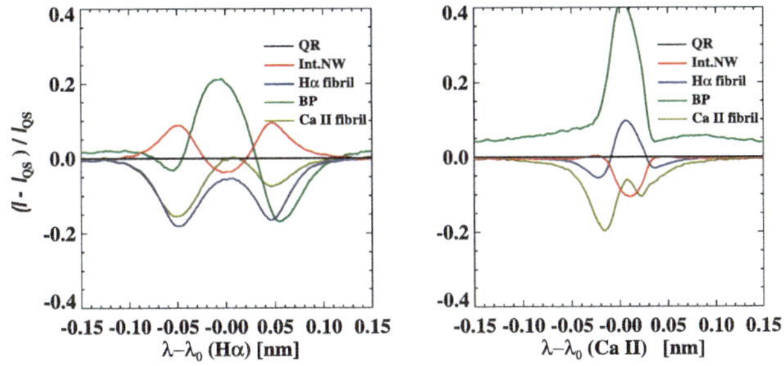

Figure 4 Mean contrast profiles of the Hα line and the Ca II 854.2 nm line for each kind of features.

and the threshold value depends on the feature. The bottom panel of Figure 3 illustrates our feature identification in the network region in a specific raster scan. The Hα fibrils occupy a significant portion of the region. The Ca II fibrils are shorter than the Hα fibrils, but a great part of these Ca II fibrils overlap the Hα fibrils, suggesting that the Ca II fibrils are not fully distinct from the Hα fibrils. The Ca II bright points partly overlap the Hα fibrils, even though they must be mutually different. The white areas refer to the unidentified regions inside the network region, and are excluded from our analysis below.

The mean contrast profiles in Figure 4 shows the spectral property of each kind of features. The left panel shows that the Hα fibrils, the most conspicuous features in Hα, appear as absorption features at all wavelengths in that they are darker than the quiet region average. The peak dark contrast of -0.17 occurs at wavelengths of about ± 50 pm. The Hα contrast profile of the Ca II fibrils is similar to the Hα fibrils, except that their dark contrast at $+50$ pm is a little lower than the Hα fibrils. The dominance of the blue wing contrast over the red wing contrast is much stronger in the Ca II line shown in the right panel, which indicates that the mean absorption profile of the Ca II fibrils is more blueshifted than that of the Hα fibrils, or blueshifted Ca II fibrils are more numerous than redshifted Ca II fibrils. Thus, supposing Hα fibrils and Ca II fibrils are not physically distinct, we conjecture that many Ca II fibrils may be identified with blueshifted Hα fibrils.

The most conspicuous features in the Ca II contrast profile plots are of course bright points. Even though the core intensity is still a minimum in most of the Ca II line profiles of the quiet region, these appear as emission features in that they are brighter than the average quiet Sun. The positive contrast is as high as 0.4 at the line center, and persists even in the wings. These bright points seen in the wings are the same as the photospheric magnetic points which appear brighter than the average quiet Sun in the continuum, G-band, UV, and broad-band Ca II H and K lines (see, *e.g.* Reardon, Uitenbroek, and Cauzzi, 2009).

4. Bisectors of Mean Spectral Profiles

4.1. The Quiet Region as a Whole

Figure 5 confirms that the bisector of the average Ca II line profile of the quiet region has an inverse C-shape. The peak of redshift of 5.1 pm – corresponding to 1.8 $\mathrm{km\,s^{-1}}$ – occurs at $I_1 = 0.27 I_c$ and $\Delta\lambda = 18$ pm, supporting the finding of Uitenbroek (2006) not only qualitatively, but also quantitatively in that the peak redshift we obtained is very close to 1.6 $\mathrm{km\,s^{-1}}$ reported by Uitenbroek (2006). Note that this average profile came from about 2.1×10^6 spectral profiles taken from both the network region and the IN, excluding some edges, and over half an hour. The observed redshift implies average downward motion in

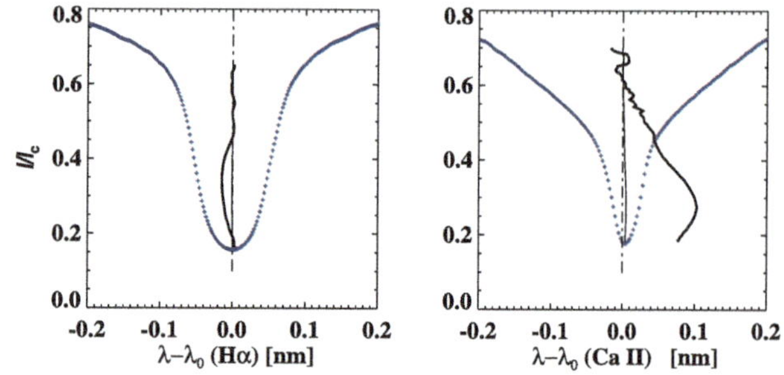

Figure 5 Enlarged presentations of the Hα line (left) and Ca II (right) profiles with the bisectors overplotted in dot-dashed curves. Thick curves represent the enhanced view of the bisectors with the wavelength offsets being magnified by a factor of 20.

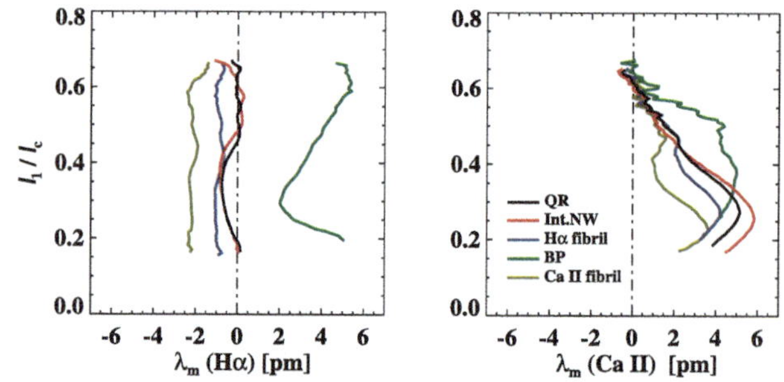

Figure 6 Plots of $I_1(\lambda_m)$ (bisectors) for each kind of features in the Hα line (left), and the Ca II line (right), respectively.

the rest frame of the solar center. We find that the upper portion of the bisector with higher values of intensity is close to the vertical line of zero Doppler shift, which is consistent with our expectation that the lower level of the atmosphere may be at rest at least in the average sense.

The bisector of the Hα line has a shape much different from that of the Ca II line. It is not of an inverse C-shape. Rather, it is almost vertically straight or, if not, of a weakly pronounced C-shape. The peak of blueshift occurs at $I_1 = 0.37I_c$ and $\Delta\lambda = 47$ pm, and has a value of − 1.0 pm – corresponding to 0.45 km s^{-1}. Interestingly, both the top and the bottom portions of the bisector are close to the vertical line of zero Doppler shift.

Our result on the bisector of the Hα line is apparently contrasted with Uitenbroek (2006) who made a brief remark that "the hydrogen Balmer α line bisector shows a hint of the pattern, with an amplitude of only 0.5 pm, corresponding to a downward motion of 228 m s^{-1}". However, this amplitude is so small that this remark might be rather interpreted as the indication that the shape of the bisector is close to a vertical line of zero shift, as our result does. The slight difference between our result and Uitenbroek (2006) might be due to the fact that the field of view in our observations is much smaller than that of Uitenbroek (2006) so that the areal fractions occupied by Hα fibrils and bright points, respectively, differ between the two studies.

4.2. Individual Features

Figure 6 presents the conventional bisectors – the plots of I_1 *vs.* λ_m – of the two lines for different kinds of features, and Figure 7 shows the plots of v_m *vs.* $\Delta\lambda$. This latter plot provides another way of displaying the spectral information contained in the bisectors. Roughly speaking, v_m for a smaller $\Delta\lambda$ reflects the line-of-sight velocity of the outer layer affecting the line, and v_m for a larger $\Delta\lambda$, that of the inner layer so that $v_m(\Delta\lambda)$ measures the variation of line-of-sight velocity over height. Note that for the construction of the average line pro-

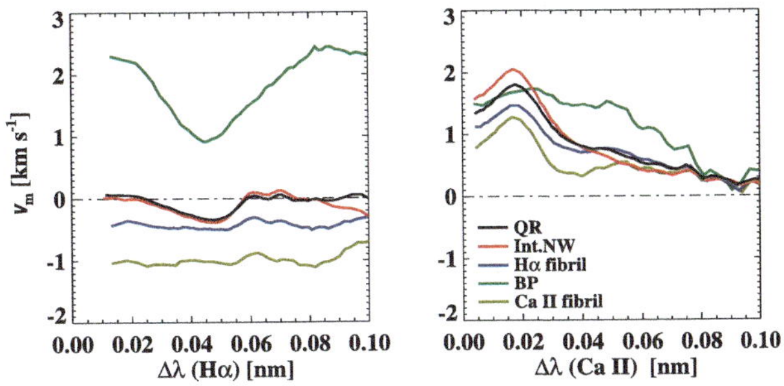

Figure 7 Plots of $v_{\rm m}(\Delta\lambda)$ for each kind of features in the Hα line (left), and the Ca II line (right), respectively.

files in these figures, we used data from 30 raster scans obtained for 9 minutes, with the total numbers of spectral profiles being about 1.2×10^5 (30 % of the examined network area) for Hα fibrils, 3.5×10^4 (9 %) for Ca II fibrils, and 1.5×10^4 (4 %) for the Ca II bright points. Meanwhile the examined internetwork area contains about 2.5×10^5 pixels, 0.63 times the network area.

Figures 6 and 7 show that IN is similar to the whole quiet region in the shapes of the Hα line bisector and the Ca II line bisector. This similarity may be attributed to the fact that IN occupies a significant fraction of the quiet region, and it does not contain conspicuous inhomogeneous features such as fibrils and network bright points. This similarity also suggests that the observed characteristics of the bisectors of the quiet region as a whole described above may mostly originate from the structure and dynamics of IN. Not surprisingly, IN is truly representative of the quiet Sun. Since IN is characterized by the absence of noticeable magnetic structures and the presence of three-minute oscillations and acoustic shocks, it appears that the property of the bisectors such as the inverse C-shape of the Ca II line is more related to the oscillations and shocks than to the magnetic geometry.

The fibrils in the network region are different from IN in the shapes of the bisectors in several aspects. First, in comparison with IN, Hα fibrils are blueshifted by about 0.3 km s^{-1} in the Hα bisector and by 0.6 km s^{-1} in the Ca II bisector, and Ca II fibrils are blueshifted by about 0.6 km s^{-1} in the Hα bisector and by 0.7 km s^{-1} in the Ca II line bisector. Second, the Hα bisectors of the fibrils have shapes different from IN, being more straight. Third, the Ca II bisector of the Ca II fibrils contains a locally straight part as well as a part of inverse C-shape, so that its shape is not so regular as that of IN. The shape of the Ca II bisector of the Hα fibrils is a hybrid between the Ca II fibrils and IN. The difference between the Hα bisectors and the Ca II bisectors of the fibrils may be simply because at places where the fibrils are visible in Hα, the Ca II line still sees the underlying chromosphere that has basically the same property as the one in IN. The fibrils do not drastically alter the Ca II line profile of the light incident from below except for the noticeable reduction in the redshift.

The network bright points are much different from other features in the shapes of the bisectors. In the Hα line, the bisector is redshifted from the reference vertical line as much as 5.5 pm or 2.5 km s^{-1}, whereas the other bisectors either remain close to the reference line or are slightly blueshifted from it (see the left panels of Figures 6 and 7). The minimum redshift of about 2 pm occurs at $I_1 = 0.29 I_{\rm c}$ and $\Delta\lambda = 47$ pm, which leads to a pronounced C-shape of the bisector. The shape of the Ca II bisector also is much different from others; neither it has a sharp redshift peak, nor the intensity of redshift peak coincides with others (see the right panels of Figures 6 and 7). Note that the large redshift of around 4 pm or 1.4 km s^{-1} persists at intensity values as high as $0.5 I_{\rm c}$, indicating that not only the core, but also the wings are redshifted. The redshift in the bright points thus seems to originate from a wide range of the atmospheric levels.

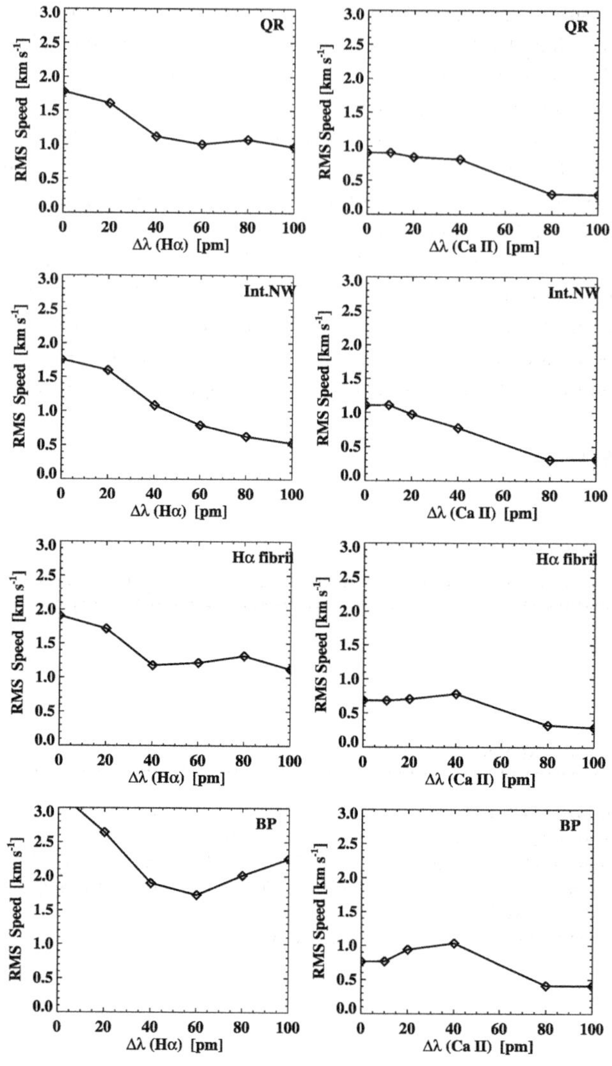

Figure 8 RMS fluctuations of v_{m} of the Hα line and Ca II 854.2 nm line, respectively, as functions of $\Delta\lambda$ for the quiet region as a whole, IN, the fibrils, and the bright points from top to bottom.

5. Spatio-Temporal Fluctuations of Doppler Shifts

Figure 8 presents the RMS Doppler shift fluctuations in the quiet region as a whole, and in each kind of identified features as functions of $\Delta\lambda$. We find that the Doppler shift fluctuation is the biggest at the line center ($\Delta\lambda = 0$) and decreases with $\Delta\lambda$, except for the bright points. This pattern in the Ca II line was previously noted by Evans and Michard (1962). It is the strongest in IN, being consistent with the picture of a stratified medium where the amplitude of velocity fluctuation increases with height to compensate for density drop.

Our measurements of Doppler shift fluctuations particularly at the line center and in the quiet region can be compared with previous measurements. We find that the RMS value is 1.8 $\mathrm{km\,s^{-1}}$ at the center of the Hα line. This value is bigger than 0.95 $\mathrm{km\,s^{-1}}$ in Table 2 of Evans and Michard (1962), but is smaller than 2.3 $\mathrm{km\,s^{-1}}$ reported by Cram, Brown, and

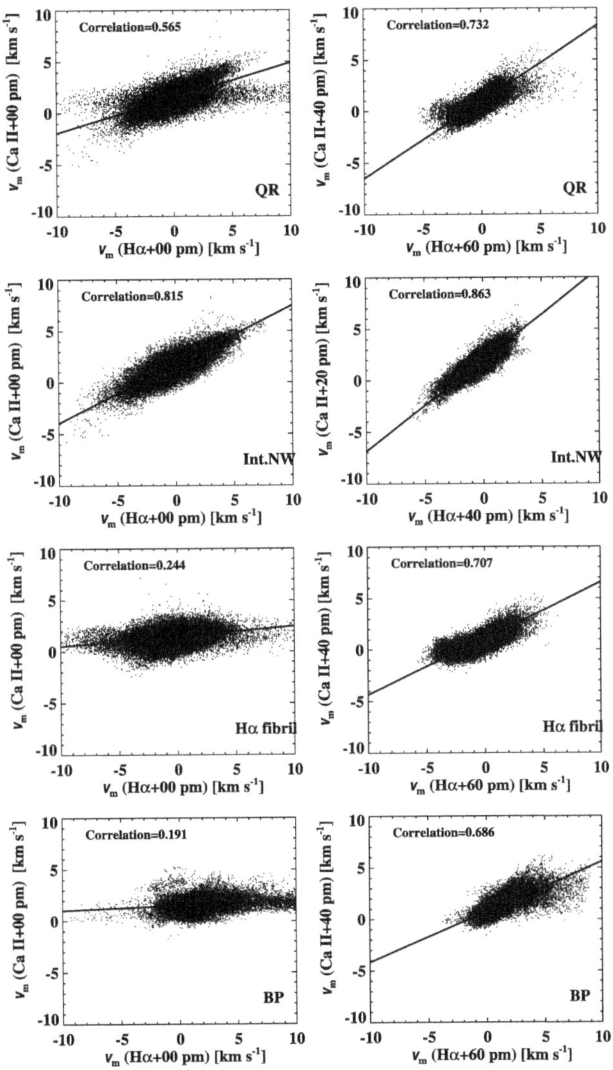

Figure 9 Scatter plots of the Hα v_m *vs.* from the Ca II v_m determined from the centers of the lines (left column) and those determined from the pair of $\Delta\lambda$ of the maximum cross-correlation (right column), for the quiet region as a whole, IN, the fibrils, and the bright points from top to bottom.

Beckers (1977). The RMS value we obtained at the Ca II 854.2 nm line center is 0.9 km s^{-1}, which is close to 0.87 km s^{-1} given in Table 3 of Evans and Michard (1962) and is smaller than 1.6 km s^{-1} reported by Cram, Brown, and Beckers (1977) and 1.4 – 1.5 km s^{-1} reported by Cauzzi *et al.* (2008). These discrepancies might be attributed either to the intrinsic region-to-region dependence of dynamical property or to the difference in the quality of seeing, as noted by Cram, Brown, and Beckers (1977).

Figure 9 presents the scatter plots showing the correlation between the two lines. We first compare between the cores of the lines and then between the two values of $\Delta\lambda$ displaying the maximum correlation. We are interested in these values because using them we can easily identify the same moving features in the two lines. They also serve as the observational basis for the understanding of how the two lines respond to the same moving features. In the quiet region as a whole, the correlation between the cores is moderate (0.57). This value is very close to 0.58, the correlation coefficient of Ca II Doppler shift and Hα Doppler shift obtained by Cauzzi *et al.* (2009), and is a little smaller than 0.63, the value obtained by Cram, Brown, and Beckers (1977). The maximum correlation of 0.73 occurs between $\Delta\lambda = 60$ pm and $\Delta\lambda = 40$ pm. In IN, the correlation between the cores is fairly high, being 0.82, and the strongest correlation of 0.86 occurs between 40 pm and 20 pm. For the Hα fibrils, the correlation is weak between the cores, and has a peak of 0.71 between 60 pm and 40 pm. For the bright points, the correlation is even weaker, having a values of 0.19 between the cores, and a peak of 0.69 between 60 pm and 40 pm.

These observed correlations of Doppler shift between the Hα and Ca II lines may be understood based on our current knowledge of the formation of these lines in the solar chromosphere. Through the cores of these lines, we see the outer layer of the chromosphere. The strong correlation between the cores in IN supports that they sense about the same outer layer of the chromosphere in IN (Cauzzi *et al.*, 2009). The weak correlation in the fibrils originate from the fact that the Hα core is much more opaque than the Ca II core and hence senses fibrils while the Ca II core does not. The weak correlation in the bright points is mainly due to the contamination of the Hα spectral data by the fibrils. When all features are considered, the core velocities plot of the quiet region as a whole clearly shows different populations. Through the wings of the lines, we see the lower atmospheric layers where fibrils are not so important. This invisibility of fibrils strengthens the correlation between the wings of the two lines even in the region of fibrils, which naturally explains the strengthening of the correlation in the quiet region as well.

The observed RMS speeds and correlations of the Doppler shift in the bright points should be interpreted with caution for a couple of reasons. The first concern is the probable inadequacy of the lambdameter method in the Ca II core in the case it appears in emission, not in absorption. Our real investigation, however, shows that there is no indication of the emission core at least in the Ca II line data taken from the network region of our interest. The other and more realistic concern is the contamination of the spectral data by the fibrils that either overlie or neighbor the bright points. This contamination may be in charge of the large RMS speed of the Hα line (Figure 8) in bright points, as well as the poor correlation of the core velocities between the two lines (Figure 9). Thus we have to admit that our data and our analysis are not much suited for the study of the Doppler shift fluctuations in the bright points. In order to extract the pure spectral information of bright points, we will need high spatial resolution data, more careful identification preventing the contamination, and a bigger sample.

Figure 10 illustrates that the major origin of Doppler shift fluctuation in IN is three-minute oscillation. Shown in the figure are the analysis of the time series of the velocities taken from a single pixel inside IN. The power spectra are peaked at discrete frequencies of 4.1, 5.2, and 7.5 mHz. The maximum power frequency is 5.2 mHz, corresponding to the period of 190 s. This result is very consistent with previous studies (Lites, Rutten, and Kalkofen, 1993; Vecchio *et al.*, 2007; Rutten, van Veelen, and Sütterlin, 2008; Vecchio, Cauzzi, and Reardon, 2009; Cauzzi *et al.*, 2009; Kontogiannis *et al.*, 2010). Our new finding about three-minute oscillations in IN is that the variation of the Hα line Doppler shift and that of the Ca II line Doppler shift are almost in phase, irrespective of whether the pair of the cores or that of the inner wings is used. We estimate the time lag between the two lines at, if any, only a few seconds, with the Hα line Doppler shift variation slightly lagging behind the Ca II line Doppler shift variation.

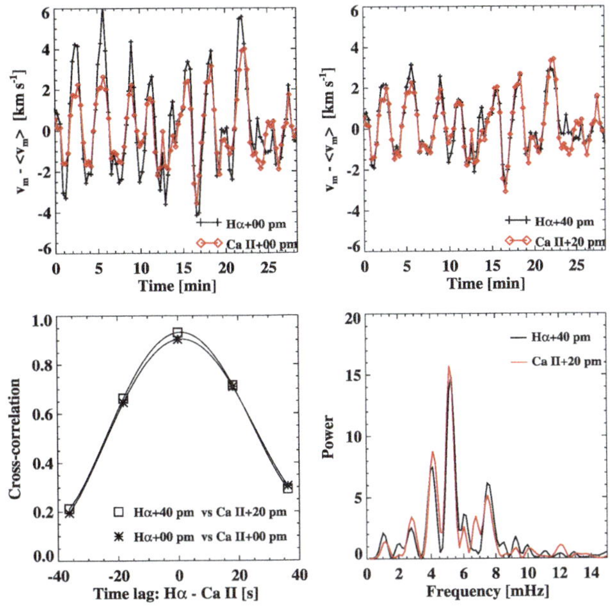

Figure 10 Top rows: temporal variations of Doppler shifts of the two lines determined at a single fixed point inside IN with different values of $\Delta\lambda$. Bottom left: the cross-correlation between the time series of Hα line Doppler shift and that of Ca II line Doppler shift. Bottom right: their power spectra.

6. Discussion

Internetwork regions are truly representative of the quiet Sun, so we first focus on the results obtained from IN inside the field of view of our observation. It is clear from our results that three-minute oscillations dominate this part and hence significantly affect the cores of these lines, and are responsible for the observed strong correlation of the Doppler shifts between the two lines, as was also remarked in previous observations (*e.g.* Rutten, van Veelen, and Sütterlin, 2008; Vecchio, Cauzzi, and Reardon, 2009). In this regard, the Hα line and the Ca II line appear to be very useful tools for the investigation of three-minute oscillations in the quiet Sun. The monotonic decrease of the RMS value with $\Delta\lambda$ in each line is quite consistent with the picture of a stratified medium with amplitude of velocity fluctuation exponentially increasing with height. The strong correlation and an insignificant phase difference between the two lines suggest that the formation heights of these lines are quite close to each other or different heights of the chromosphere oscillate in phase. Since the RMS values of the Ca II Doppler shifts are a little smaller than those of the Hα line, the average formation height of the Ca II line is likely to be a little lower than the Hα. From Figure 8, we estimate the RMS ratio of the Ca II line to the Hα line at $[0.6, 0.9]$. Supposing that these ratios are equal to $\exp(\Delta z/2H_p)$ where Δz is the difference in the formation height between the two lines, and H_p is pressure scale height, we obtain $\Delta z = [0.2, 1.0]H_p \approx [40, 200]$ km.

Since the Hα line and the Ca II line do not differ much in the formation height, the observed contrast between these lines in the bisector shape of the mean profiles may be attributed to the different responses to the structure and dynamics of the outer chromosphere. The Ca II lines are much affected by collisional processes, and hence reflect the temperature of the line formation region. In contrast, the Hα line is dominated by radiative processes such as radiative excitation/deexcitation and photoionization/radiative recombination, and hence is insensitive to the local kinetic temperature in the chromosphere (*e.g.* Cauzzi *et al.*, 2009). The redshifted Ca II line bisector of an inverse C-shape suggests that for some reason, the Ca II line selects a little more of downflows than upflows while the Hα line does not.

The formation of the Ca II 854.2 nm line was studied by Uitenbroek (2006) with the non-LTE radiative transfer computations based on a snapshot from a simulation of solar convection (Asplund *et al.*, 2000). Leenaarts *et al.* (2009) computed synthetic Ca II 854.2 nm images by applying a three-dimensional non-LTE radiative transfer code to a radiation-MHD simulation of the solar atmosphere. These computations successfully reproduced various features seen in the blue wing and in the core, respectively. But they were not successful in reproducing the inverse C-shape of the observations. The spatially average profile computed by Uitenbroek (2006) was redshifted up to 0.5 $\mathrm{km\,s^{-1}}$ with respect to the central wavelength, but not so much as observed. In addition, both simulations (Uitenbroek, 2006; Leenaarts *et al.*, 2009) failed to produce the observed profiles especially near the cores; the simulated cores were much narrower and deeper than the observed ones. Leenaarts *et al.* (2009) attributed this discrepancy to the existence of motion in small scales of 0.1 Mm that future simulations with higher spatial resolution can resolve. The understanding of the Hα core formation in the chromosphere is more difficult because this line is strongly scattering and is much subject to non-local effects. Leenaarts, Carlsson, and Rouppe van der Voort (2012) confirmed that these non-local or smoothing effects are crucial for explaining the features observed through the core of this line; their 3D computation successfully reproduced predominantly dark, fibril-like features while their 1D computations failed to do that, only showing a distinctly photospheric scene.

Our results indicate that the simultaneous recording of the Hα line and the Ca II 854.2 nm line is a very useful diagnostic tool for the investigation of three-minute oscillations in IN. Note that the proper use of this tool, however, requires a correct knowledge on how these features affect the observed profiles of these lines. For example, Doppler velocities inferred from the line profiles using the lambdameter method may differ much from the true plasma velocities in the upper chromosphere, even though the two kinds of velocities are expected to be strongly correlated with each other. It should be kept in mind that the line is formed by the lower parts of the atmosphere as well. Therefore, the Doppler shift of the line should not be identified with the line-of-sight velocity of the outer part, but with a weighted average of the whole atmosphere. Generally speaking, since the lower parts of the atmosphere are likely to have lower line-of-sight velocities than the outer part, the velocity of the outer part may be higher than the inferred Doppler shift.

We plan to develop a simple model of radiative transfer in order to extract the Doppler velocity, Doppler width and other parameters of the outer part only from the observed line profile. A well-known model that has been widely used for a similar purpose was the cloud model developed by Beckers (1964). This cloud model can be successfully applied to light-scattering features lying high above other chromospheric features, but not to features embedded in the chromosphere such as waves and shocks. With a new cloud model applicable to the embedded features being available, we will be able to determine the Doppler widths in waves and shocks in IN and hence will be able to infer their temperatures. Note that this simple model is for the inversion only and is not like, for example, the forward physical model of Leenaarts, Carlsson, and Rouppe van der Voort (2012) that reproduces fibril-like structures by combining radiation-MHD simulations and full non-LTE radiative transfer computations. We also plan to improve the imaging quality of the FISS observations based on the upgrade of the AO and the relay optics. With these two things done, we will able to fully exploit the capability of the simultaneous recording of the Hα line and Ca II 854.2 nm line to reveal the origin of asymmetries in these lines as well as for the study of three-minute oscillations in IN and other features in network regions.

Acknowledgements We are grateful to the referee for a number of constructive comments. This work was supported by the National Research Foundation of Korea (2011-0028102) and by the Development of Korean Space Weather Center, a project of KASI.

References

Asplund, M., Nordlund, Å., Trampedach, R., Allende Prieto, C., Stein, R.F.: 2000, *Astron. Astrophys.* **359**, 729.
Beckers, J.M.: 1964, A Study of the Fine Structures in the Solar Chromosphere. Ph.D. Thesis, University of Utrecht.
Cauzzi, G., Reardon, K.P., Uitenbroek, H., Cavallini, F., Falchi, A., Falciani, R., Janssen, K., Rimmele, T., Vecchio, A., Wöger, F.: 2008, *Astron. Astrophys.* **480**, 515.
Cauzzi, G., Reardon, K., Rutten, R.J., Tritschler, A., Uitenbroek, H.: 2009, *Astron. Astrophys.* **503**, 577.
Chae, J., Park, H.-M., Ahn, K., Yang, H., Park, Y.-D., Nah, J., Jang, B., Cho, K.-S., Cao, W., Goode, P.R.: 2012, *Solar Phys.* doi:10.1007/s11207-012-0147-x.
Cram, L.E., Brown, D.R., Beckers, J.M.: 1977, *Astron. Astrophys.* **57**, 211.
Deubner, F.-L.: 1974, *Solar Phys.* **39**, 31.
Deubner, F.-L., Fleck, B.: 1990, *Astron. Astrophys.* **228**, 506.
Deubner, F.-L., Waldschik, T., Steffens, S.: 1996, *Astron. Astrophys.* **307**, 936.
Evans, J.W., Michard, R.: 1962, *Astrophys. J.* **135**, 812.
Gray, D.F.: 1992, *The Observation and Analysis of Stellar Photospheres*, Cambridge University Press, Cambridge, 417.
Kontogiannis, I., Tsiropoula, G., Tziotziou, K., Georgoulis, M.K.: 2010, *Astron. Astrophys.* **524**, A12.
Leenaarts, J., Carlsson, M., Rouppe van der Voort, L.: 2012, *Astrophys. J.* **749**, 136.
Leenaarts, J., Rutten, R.J., Sütterlin, P., Carlsson, M., Uitenbroek, H.: 2006a, *Astron. Astrophys.* **449**, 1209.
Leenaarts, J., Rutten, R.J., Carlsson, M., Uitenbroek, H.: 2006b, *Astron. Astrophys.* **452**, L15.
Leenaarts, J., Carlsson, M., Hansteen, V., Rouppe van der Voort, L.: 2009, *Astrophys. J. Lett.* **694**, L128.
Leighton, R.B., Noyes, R.W., Simon, G.W.: 1962, *Astrophys. J.* **135**, 474.
Lites, B.W., Rutten, R.J., Kalkofen, W.: 1993, *Astrophys. J.* **414**, 345.
Mein, N., Mein, P.: 1976, *Solar Phys.* **49**, 231.
Reardon, K.P., Uitenbroek, H., Cauzzi, G.: 2009, *Astron. Astrophys.* **500**, 1239.
Rouppe van der Voort, L., Leenaarts, J., de Pontieu, B., Carlsson, M., Vissers, G.: 2009, *Astrophys. J.* **705**, 272.
Rutten, R.J.: 2010, *Mem. Soc. Astron. Ital.* **81**, 565.
Rutten, R.J., van Veelen, B., Sütterlin, P.: 2008, *Solar Phys.* **251**, 533.
Sekse, D.H., Rouppe van der Voort, L., De Pontieu, B.: 2012, *Astrophys. J.* **752**, 108.
Scoas-Navarro, H., Elmore, D.: 2005, *Astrophys. J. Lett.* **619**, L195.
Uitenbroek, H.: 2006, *Astrophys. J.* **639**, 516.
Vecchio, A., Cauzzi, G., Reardon, K.P.: 2009, *Astron. Astrophys.* **494**, 269.
Vecchio, A., Cauzzi, G., Reardon, K.P., Janssen, K., Rimmele, T.: 2007, *Astron. Astrophys.* **461**, 1.

DOI 10.1007/978-3-319-12123-9_7
Reprinted from *Solar Physics* Journal, DOI 10.1007/s11207-013-0271-2

Temperature of Solar Prominences Obtained with the Fast Imaging Solar Spectrograph on the 1.6 m New Solar Telescope at the Big Bear Solar Observatory

Hyungmin Park · Jongchul Chae · Donguk Song · Ram Ajor Maurya · Heesu Yang · Young-Deuk Park · Bi-Ho Jang · Jakyoung Nah · Kyung-Suk Cho · Yeon-Han Kim · Kwangsu Ahn · Wenda Cao · Philip R. Goode

Received: 2 July 2012 / Accepted: 6 March 2013 / Published online: 29 March 2013
© Springer Science+Business Media Dordrecht 2013

Abstract We observed solar prominences with the Fast Imaging Solar Spectrograph (FISS) at the Big Bear Solar Observatory on 30 June 2010 and 15 August 2011. To determine the temperature of the prominence material, we applied a nonlinear least-squares fitting of the radiative transfer model. From the Doppler broadening of the Hα and Ca II lines, we determined the temperature and nonthermal velocity separately. The ranges of temperature and nonthermal velocity were 4000 – 20 000 K and 4 – 11 km s^{-1}. We also found that the temperature varied much from point to point within one prominence.

Keywords Chromosphere, quiet · Heating, chromospheric · Prominences, quiescent · Spectral line, broadening

1. Introduction

A proper understanding of the nature of solar prominences relies on precise measurements of their basic physical parameters: temperature, density, mass motion, and magnetic field. Temperature is of particular interest in this study. It is well known that the temperature of the prominence material is about 10^4 K, whereas the corona has a temperature of 10^6 K. Precise measurements of the temperature in a prominence are needed to estimate the other physical parameters. For example, hydrogen is partially ionized in the prominence so that

Initial Results from FISS
Guest Editor: Jongchul Chae

H. Park (✉) · J. Chae · D. Song · R.A. Maurya · H. Yang
Astronomy Program, Department of Physics and Astronomy, Seoul National University, Seoul, Republic of Korea
e-mail: hmpark@astro.snu.ac.kr

Y.-D. Park · B.-H. Jang · J. Nah · K.-S. Cho · Y.-H. Kim
Korea Astronomy and Space Science Institute, Daejeon, Republic of Korea

K. Ahn · W. Cao · P.R. Goode
Big Bear Solar Observatory, New Jersey Institute of Technology, Big Bear City, CA, USA

the degree of ionization may be sensitive to the temperature. In addition, the density can only be determined if the temperature is known.

The most common way to determine the prominence temperature is to use the absorption width of a spectral line. The major broadening in the line is the Doppler broadening, which contains the thermal as well as the nonthermal contribution. If the value of the nonthermal contribution is known, one can obtain the thermal contribution from the line broadening. In some studies, the nonthermal contribution was neglected so that the obtained temperatures represented the upper limit to the real value (Ellison, 1952; Jefferies, 1955; Mein and Mein, 1991). In other studies, more than one line was used, and both the thermal and nonthermal contributions were determined independently (Hirayama, 1963, 1971; Jefferies, 1956, 1962; Gallegos and Machado, 1973; Nikolsky, Gulyaev, and Nikolskaya, 1971; Landman, Edberg, and Laney, 1977; Zhang *et al.*, 1987; Li *et al.*, 1998; Mein and Mein, 1991; Stellmacher, Wiehr, and Dammasch, 2003).

Temperatures obtained through the line width method have a wide range from 4000 K to 20 000 K. How can we understand these results? The temperature probably depends on the individual prominences observed. Another possible explanation is that it depends on the spectral lines used. For instance, Hirayama (1963) obtained a prominence temperature of 4000 K from helium and metallic lines, and Jefferies (1956) obtained 20 000 K by combining the Hα line with the helium D3 line. It is not surprising that the measured temperatures depend on the line used. The visibility of a line from a specific atom is determined by the ionization and excitation process in the prominence, which depends on the temperature. Therefore, each detected line may have its own range of temperature. Finally, some of the previous studies indicated that the temperature varies with the position and time inside each prominence. To understand the wide range of the reported temperatures of the prominences, more observations are needed, especially using spectrographs with space- and time-resolving capabilities.

The Fast Imaging Solar Spectrograph (FISS) is one of the instruments of the 1.6 m New Solar Telescope (NST) at the Big Bear Solar Observatory. It is an imaging spectrograph designed to produce data with high spectral, spatial, and temporal resolutions that are suited for the study of fine-scaled structures and dynamics in the solar chromosphere (Chae *et al.*, 2012). The FISS has the capability of recording two spectral bands simultaneously and performing imaging based on the linear motion of a two-mirror scanner. From the Hα and the Ca II 8542 band, one can measure the temperature and nonthermal velocity in chromospheric features such as filaments (prominences), quiet regions, and active regions. In addition, the imaging capability of the FISS enables us to study the dynamics of such features. These capabilities can also be exploited for observations of solar prominences. So far, researchers have mostly observed only a limited part of a prominence with a spectrograph; our observations of prominences with the FISS will provide detailed knowledge on the variation with position of the physical quantities within individual prominences.

We here describe prominence observations using the FISS with our data reduction and spectral analysis, and the results on the temperatures of solar prominences. For the spectral analysis, we employed a simple model of radiative transfer that takes into account the effect of finite optical depth.

2. Data and Analysis

2.1. Observation

The observations were performed with the FISS on the 1.6 m New Solar Telescope (NST) at the Big Bear Solar Observatory (BBSO). This instrument is a high-dispersion Echelle spec-

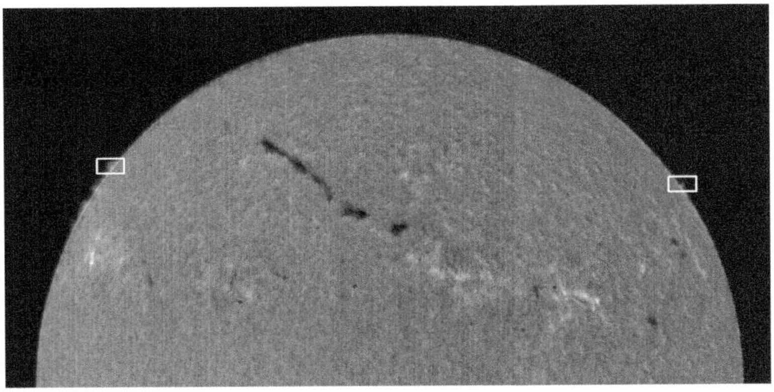

Figure 1 Hα image obtained on 30 June 2010 at the BBSO. The two rectangles represent the prominences on the east and west limb observed by the FISS. The field-of-view is $72'' \times 40''$.

trograph of near-Littrow type, with a 32 μm slit width ($0.16''$), a focal length of 1.5 m, a focal ratio of F/26, and high spectral orders: 34th at Hα and 26th at Ca II 8542 Å. It is currently the largest operating solar telescope in the world with Echelle grating and fast CCD, and we obtained data with high spatial, spectral, and time resolutions. More information about the FISS performance can be found in Chae *et al.* (2012).

On 30 June 2010, we observed one prominence on the east limb and another on the west limb (Figure 1). The two rectangles in the figure show the two fields of view of the observations. The exposure time was 90 ms, and the scanning time was 140 s. In addition, we observed a prominence in the east limb on 15 August 2011. In contrast to 30 October 2010, the adaptive optics (AO) system worked on this day, and the prominence was scanned stably with little image motion along the slit direction. In this observation, the exposure was 30 ms, and the scanning time was 40 s. The observed spectra were treated with basic processes such as dark-subtraction and flat-fielding. The processed data were compressed and stored based on the principal component analysis (PCA). This PCA compression reduces noise as well as the data size without much loss of information (Chae *et al.*, 2012). In addition, the aureole light spectra on the Hα and Ca II lines were removed carefully by subtracting the spectra from outside the prominence. The two spectral bands were spatially aligned by shifting the Ca II spectra by a fixed amount with respect to the Hα spectra.

2.2. Spectral Analysis

The equation for radiative transfer across a layer is well known. The equation consists of background radiation and the contribution of the source from the object itself. In the case of limb prominences, background radiation can be ignored. If the source function is constant along the line of sight, the intensity is theoretically modeled with the equation using optical thickness τ

$$I = S\left(1 - \mathrm{e}^{-\tau}\right), \tag{1}$$

where S is the source function.

In prominences/filaments, thermal broadening is much more significant in the formation of spectral lines than collisional broadening, so that the optical thickness profile may be written in the Gaussian functional form

$$\tau = \tau_0 \exp\left[-\left(\frac{\lambda - \lambda_c}{\Delta\lambda_D}\right)^2\right], \tag{2}$$

where $\Delta\lambda_D$ is the Doppler width, τ_0 is the optical thickness at line center, and λ_c is the central wavelength of the absorption profile, which is a measure of the Doppler shift.

It is well known that Doppler broadening contains a thermal and a nonthermal contribution. If we also assume that the temperature is constant along the line of sight and the velocity distribution of the nonthermal motion is Gaussian as well, the Doppler width in Equation (2) is expressed by

$$\Delta\lambda_D = \frac{\lambda}{c}\sqrt{\frac{2kT}{M} + \xi^2}, \tag{3}$$

where λ is the wavelength at rest, T is the mean kinetic temperature along the line of sight, M is the mass of the atom, and ξ is the nonthermal velocity (Tandberg-Hanssen, 1995; Labrosse *et al.*, 2010). The nonthermal velocity ξ incorporates all kinds of unresolved motion along the line of sight, including random motions of fine-structure threads as investigated by Gunár *et al.* (2012). We determined the values of these parameters with a least-squares fitting.

We applied the radiative transfer model to the spectra of the Hα and Ca II 8542 Å lines. Three methods were tested to estimate the temperature of these prominences. First, we applied the model to each line separately and obtained two Doppler-broadening values ($\Delta\lambda_{D,H\alpha}$, $\Delta\lambda_{D,Ca}$). Second, we regarded the temperature and nonthermal velocity as independent parameters and fitted the two line profiles together. Finally, we adopted the second approach above, but reduced the number of free parameters by introducing the line-of-sight velocity as an independent variable instead of the center positions of the two lines. This reduction is based on the assumption that the two lines are formed in the same volume.

We found that the results are almost the same and depend very little on the chosen method. We decided to choose the first method to determine the prominence temperatures since it is commonly used, can be applied to each line separately, and needs fewer assumptions.

In the analysis, it is not necessary to take into account the instrumental broadening. Using a HeNe (632.8 nm) laser, Chae *et al.* (2012) estimated the instrumental broadening as 1.6 pm, which corresponds to about 2.0 pm at the wavelengths of Hα and Ca II 8542 Å, which is so small that it is negligible.

3. Results

Figure 2 indicates how a prominence on the limb appears at the centers of the Hα (top row) and Ca II (bottom row) lines without AO. The prominence appears to be much darker than the disk in the Ca II line, while its brightness is comparable to the disk value in the Hα line. We found that the intensity gradually decreases from the disk to the prominence in the Hα line, but sharply drops in the Ca II line. Despite this difference, the Hα and Ca II images of the prominence look similar. There are some differences in the morphology between these lines, which may not be physically fundamental, but seem to reflect mostly the difference in optical thickness, which will be described below. From the Hα and Ca II spectra, we found that the Ca II line width is smaller than the Hα line width. Moreover, the solar limb appears to be irregular, which is due to the image motion from atmospheric disturbances. Since a prominence is fainter than the disk and not visible in molecular bands such as TiO and G bands, it will be hard to observe the prominence with the AO properly operating. If a prominence happens to lie near a sunspot as it did on 15 August 2011, it is possible to obtain the data with better quality because the AO can operate.

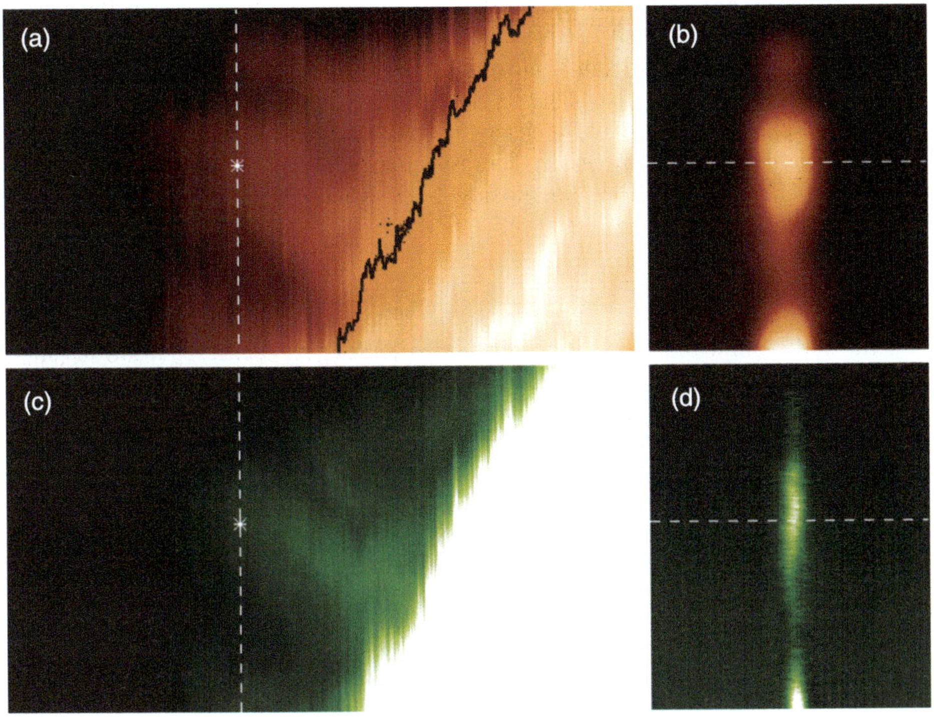

Figure 2 Scan images (left column) and spectrograms (right column) of the prominence on the east limb observed in Hα (top row) and Ca II 8542 Å (bottom row) on 30 June 2010. A black curve in the upper left image marks the boundary of the solar disk in Ca II as shown in the lower left image. Vertical dashed lines in the left-hand-side panels represent the slit position along which the spectrograms in the right-hand-side panels were taken. Horizontal dashed lines in the right-hand-side panels correspond to the positions indicated by asterisks in the left-hand-side panels. Line profiles along the horizontal dashed lines are shown in Figure 3.

To illustrate the spectral properties of the prominence in the Hα and Ca II lines, we selected an arbitrary position (marked by $*$) as indicated in Figure 2. Figure 3 shows the line profiles and fitting results in the Hα and Ca II lines. The Hα line was fitted with $\tau_0 = 3.43$, $\Delta\lambda_0 = 0.034$ nm, and $V_{\mathrm{D}} = -1.15\ \mathrm{km\,s^{-1}}$. The Ca II line was fitted with $\tau_0 = 0.29$, $\Delta\lambda_0 = 0.023$ nm, and $V_{\mathrm{D}} = -2.12\ \mathrm{km\,s^{-1}}$. The Hα line profile has a higher peak value than the Ca II line profile. Furthermore, the signal-to-noise ratio in the Hα line profile is higher than that of the Ca II. Despite this difference, both lines are fairly well fitted. From the model fitting, we found that the prominence is optically thick in the Hα line, but optically thin in the Ca II line. In an optically thin case, optical thickness and source function cannot be separately determined; therefore, their values should be taken carefully. The Doppler velocity in the Hα line is found to be close to the Ca II line, which supports our assumption that the two lines are formed in the same volume of the prominence. As a result, the temperature and the nonthermal velocity are estimated to be about 11 000 K and 7.8 $\mathrm{km\,s^{-1}}$. These values are not far from previously reported values.

The left panel of Figure 4 presents the scatter plot of the Doppler width ($\Delta\lambda_{\mathrm{D,H\alpha}}$) *versus* the Doppler width($\Delta\lambda_{\mathrm{D,Ca}}$) for the east-limb prominence of 30 June 2010. The Doppler widths are in the range of 0.028 – 0.037nm in the Hα line and 0.012 – 0.031nm for the Ca II line. The correlation coefficient (C) between the line width is 0.61. Blue and red curves, obtained from Equation (3), represent the iso-T and iso-ξ curves, respectively. From these

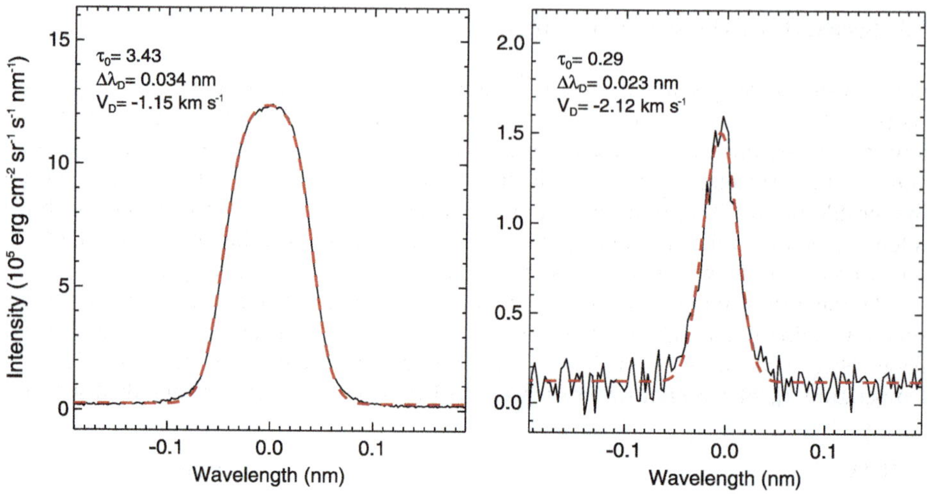

Figure 3 Line profile and fitting result for Hα (left) and Ca II (right). These two profiles come from the position of the asterisk symbols in Figure 2. Solid (black) and dashed (red) lines are the line profiles and the fitting result from our analysis.

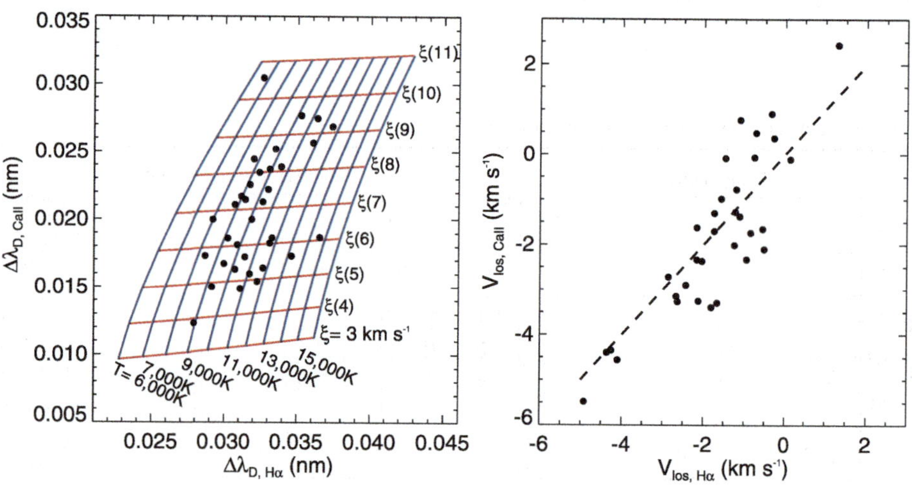

Figure 4 Scatter plots between $\Delta\lambda_{\mathrm{D,H\alpha}}$ and $\Delta\lambda_{\mathrm{D,Ca}}$ (left panel), and between $V_{\mathrm{los,H\alpha}}$ and $V_{\mathrm{los,Ca}}$ (right panel) for the east-limb prominence on 30 June 2010. In the left-hand-side panel, the blue and red curves represent iso-temperature and iso-nonthermal velocity curves. In the right-hand-side panel, the dashed line represents the same Doppler velocities in both lines.

curves, we can estimate the temperature and nonthermal velocity for each measurement marked by a dot. We found that T ranges from 7000 K to 14 000 K, and ξ ranges from 4 km s^{-1}to 11 km s^{-1}. The most probable values for the temperature and nonthermal velocity are estimated to be $10\,200 \pm 1580$ K and 6.9 ± 1.6 km s^{-1}.

The right panel shows the scatter plot between the Doppler velocity $V_{\mathrm{los,H\alpha}}$ and $V_{\mathrm{los,Ca}}$. If the Hα and Ca II lines are formed in the same volume, the Doppler velocities derived from the two lines are expected to coincide, which leads to the theoretical relationship indicated

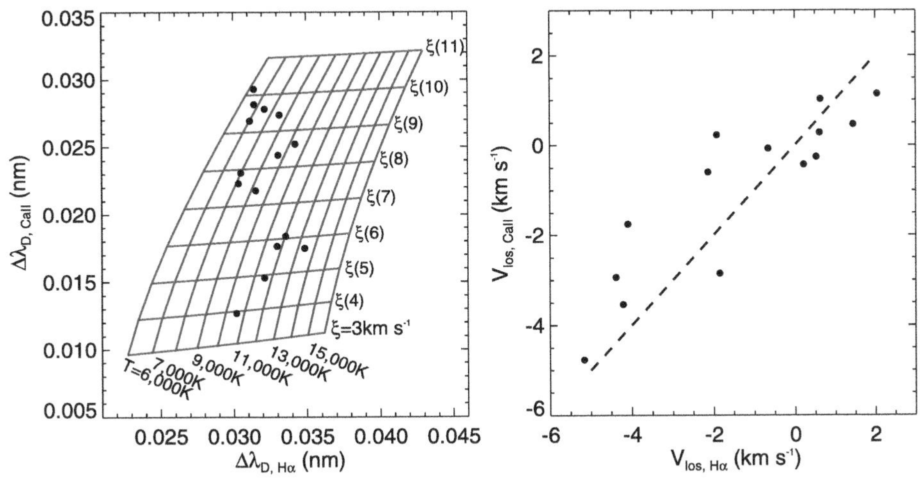

Figure 5 The same as Figure 4, but for the west-limb prominence observed on 30 June 2010.

by the dashed line (right panel). We find from the plot that the measured Doppler velocities do not deviate much from the theoretical values. The standard deviation from this line is $0.9\ \mathrm{km\,s^{-1}}$, which we consider to be tolerable.

Figure 5 presents the results from the west-limb prominence of 30 June 2010. Since this prominence is smaller and fainter than the eastern one, there are fewer data points. The measured ranges of temperature of $6000\ \mathrm{K} < T < 14\,000\ \mathrm{K}$ and the nonthermal velocity of $4\ \mathrm{km\,s^{-1}} < \xi < 10\ \mathrm{km\,s^{-1}}$ are not much different from the east-limb prominence. The correlation coefficient between the line width is $C = -0.05$. The most probable values for the temperature and nonthermal velocity are estimated to be $T = 9390 \pm 2170\ \mathrm{K}$ and $\xi = 7.6 \pm 1.9\ \mathrm{km\,s^{-1}}$. The standard deviation (right panel) of the difference between the measure and theoretical values of the nonthermal velocity is found to be $1.17\ \mathrm{km\,s^{-1}}$, which is a little higher than the above value, but still is fairly acceptable.

Figure 6 shows the scatter plot of the same parameters for the prominence on 15 August 2011. Doppler widths for the Hα and Ca II lines are found in the ranges 0.025 – 0.031 nm and 0.01–0.021 nm, respectively. The correlation coefficient between the line width is $C = 0.63$. The most probable values for the temperature and nonthermal velocity are estimated to be $T = 8680 \pm 1070\ \mathrm{K}$ and $\xi = 4.6 \pm 0.9\ \mathrm{km\,s^{-1}}$. In addition, the standard deviation of the Doppler velocities from the theoretical values is estimated to be $0.38\ \mathrm{km\,s^{-1}}$. The Doppler widths and standard deviation for the prominence on 15 August 2011 are smaller compared to the two prominences on 30 June 2010. We can explain these differences in terms of the operation of the adaptive optics (AO) system, short exposure time, or the characteristics of individual prominences. A better operation of the AO and a short exposure enable us to obtain data with higher spatial and temporal resolutions. The operation of the AO corrects for atmospheric distortion, which allows us to obtain data with higher spatial resolution and smaller errors in data alignment. The short exposure may also be useful in freezing the seeing. On the other hand, a short exposure may also bring about another effect of obscuring faint parts, with bias given to bright parts. Alternatively, the difference between this prominence and the other prominences might be due to intrinsic variations from prominence to prominence.

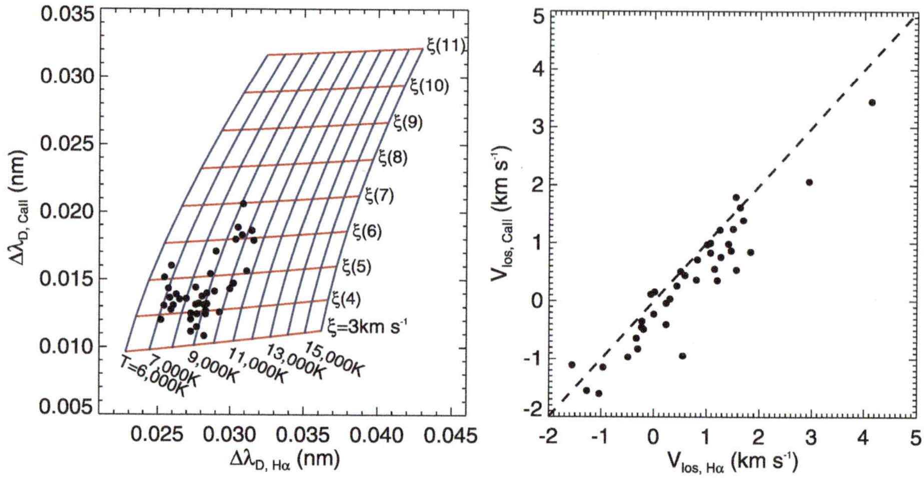

Figure 6 The same as Figure 4, but for the west-limb prominence on 15 August 2011.

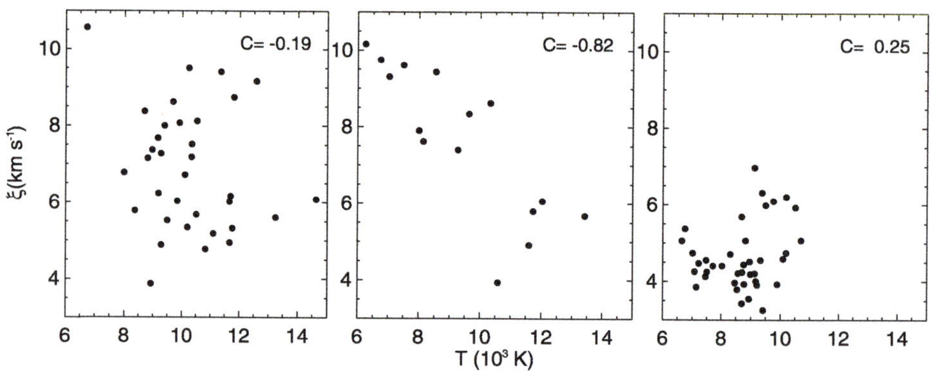

Figure 7 Scatter plots of T *versus* ξ of the east-limb prominence of 30 June 2010 (left), the west-limb prominence of 30 June 2010 (middle), and the east-limb prominence of 15 August 2011 (right). The correlation coefficient between T and ξ of the prominences is shown in each panel.

Figure 7 is presented to investigate the correlation between the temperature and nonthermal velocity in the three prominences. The left (east-limb prominence of 30 June 2010) and the right (east-limb prominence of 15 August 2011) panels show that the temperature and nonthermal velocity are not correlated ($C = -0.19$ and $C = 0.25$, respectively). In the middle panel (west-limb prominence of 30 June 2010), a negative correlation ($C = -0.82$) exists between the temperature and nonthermal velocity. Since the prominence is small and the data quality is poorer than that of the other data, it is hard to accept this result. We have checked whether the model fitting is appropriate for the prominence data and whether we can trust our results or not. We conclude that this correlation is due to the relationship between the temperature and nonthermal velocity as indicated by Equation (3). If the broadening is constant and the temperature is high, the nonthermal velocity should have a low velocity, and vice versa. These values may affect the results shown in the middle panel of Figure 7.

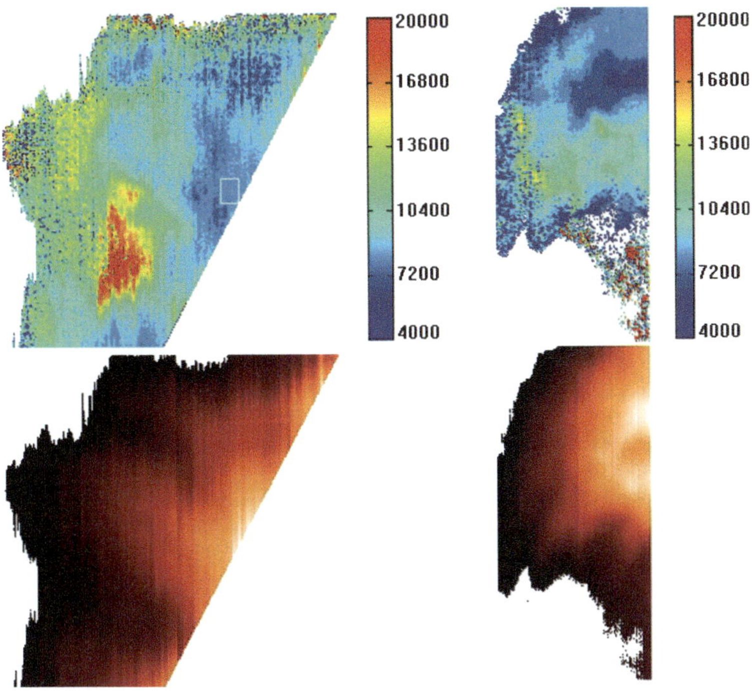

Figure 8 The temperature (top row) and intensity maps (bottom row) of the prominences observed on 30 June 2010 (left column) and on 15 August 2011 (right column). A white square marked in the upper left corner represents the region where the random error in the temperature measurement is estimated. The temperature maps are derived from the model fitting, the intensity maps from the integrated intensities.

Figure 8 presents maps of the temperature and integrated intensity of the two selected prominences of 30 June 2010 (left column) and 15 August 2011 (right column). The temperature ranges from about 4000 K to about 20 000 K in the two prominences. Obviously, the temperature varies from region to region within the same prominence. A particular example is the region with the highest temperature ($T \approx 20\,000$ K) seen in the prominence of 30 June 2010. Interestingly, this region corresponds to a fainter part of the prominence. Excluding this feature, we see a tendency that the temperature increases outward in the prominence. This kind of spatial variation was previously reported by Hirayama (1971). Conversely, this tendency is not obvious in the other prominence observed on 15 August 2011. The temperature in the boundary is comparable to the central part of the prominence. This is somewhat different from the conventional expectation that the outer part will have a higher temperature because it is in contact with the hot corona (the so-called prominence-corona transition region – PCTR). We failed in finding any indication of a temperature increase at the fainter outer parts. Currently, we have no reasonable explanation for this peculiar behavior. This prominence appears to be consistent with the result of Zhang *et al.* (1987) and Mouradian and Leroy (1977) in that there is no regular increase in temperature from the center to the edge in a prominence. As a result, our study presents examples of both patterns.

We measured the random error in the temperature of the prominence. As marked in Figure 8, we selected a region with an apparently uniform temperature and determined the standard deviation in the region of interest as a measurement of random error in the tem-

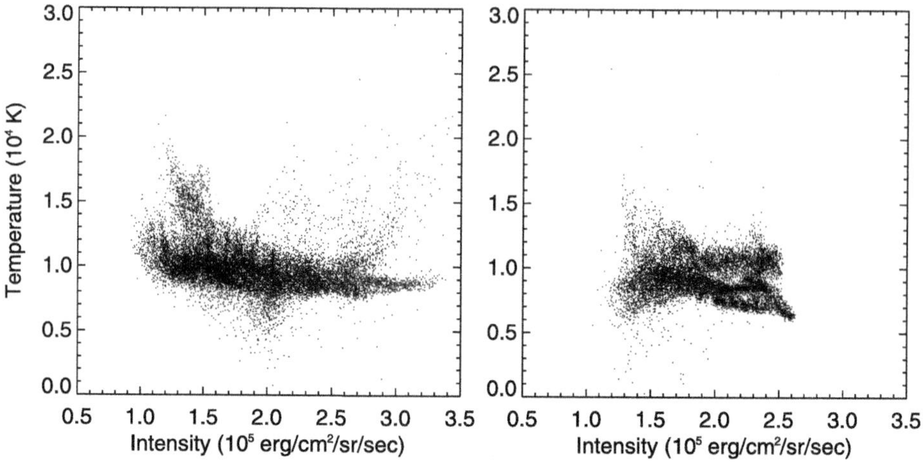

Figure 9 Scatter plots of T *versus* I (integrated intensity over the line) for two selected prominences on 30 June 2010 (left panel) and on 15 August 2011 (right panel). The estimated temperatures and correlation coefficients of the prominences are $10\,600 \pm 2440$ K and 9450 ± 2000 K, and -0.31 and -0.15, respectively.

perature. The error was estimated to be less than 300 K. Compared with the temperature we obtained, the error is small and tolerable.

Figure 9 presents the scatter plots between the integrated intensity over the line and the temperature of the two prominences observed on 30 June 2010 (left panel) and 15 August 2011 (right panel). It shows that the temperatures in the prominences are centered around 10 000 K. The prominence temperature decreases with the intensity, and the correlation coefficients between their temperatures and intensities are determined to be -0.31 and -0.15, respectively. We found a similar pattern in that the spread in the temperature is smaller within the pixels of higher intensities than within the pixels of lower intensities. This tendency may be real because the spreads are much larger than the estimated measurement errors.

Our analysis has taken into account the effects of finite optical thickness. If the optical thickness is larger than unity, the core will saturate to a constant value so that the line profile appears to be more broadened. Unless this effect is properly taken into account, the Doppler width will be overestimated, and hence the temperature. Our analysis is free from this problem since the optical thickness was treated as an independent parameter. Figure 10 shows the results of the fitting to the Hα line profiles at two different positions in the prominence. The fitting in the left panel yields 3.5 as the estimate for the optical thickness. If the optical thickness is set to either a lower or higher value, the fitting becomes poorer, as can be seen from the figure (red and green dashed curves). This supports the validity of our treatment for the effect of finite optical thickness. The right panel, however, provides an example of a case where our analysis is not satisfactory. The line profile in this case shows a central reversal in the core, which may be due to self-absorption arising when the optical thickness is very large. The failure of our analysis in this case comes from the assumption that the source function is constant along the line of sight. This is a limitation of the model. Nevertheless, there are so few data points showing the central reversal in the Hα line that this limitation does not significantly affect the statistics of the temperature measurements.

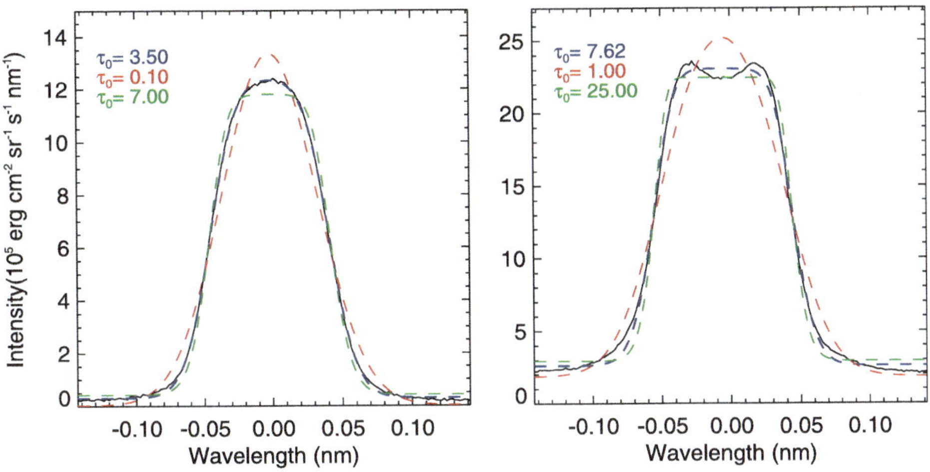

Figure 10 Hα line profiles and fitting results with different τ_0 values at two different positions on the prominence of 30 June 2010. The black lines represent Hα line profiles. The blue dashed curve represents the fitted profile with optical thickness τ_0 while the red dashed and green dashed curves represent the variation in the fitted profiles due to small/large values in the optical thickness from the fitted τ_0 over the Hα line.

4. Discussion

In three prominences we determined the temperature of Hα-emitting plasma based on the simultaneous measurements of the Doppler widths of the emission profile in the Hα and Ca II 8542 Å lines. Our study showed that the temperature within each prominence varies from 4000 to 20 000 K and the nonthermal velocity varies from 4 to 11 km s^{-1}. The temperature has no correlation with the nonthermal velocity, but has a weak negative correlation with the brightness.

The estimated temperature range in the prominences is broad enough to be consistent with previous reports. According to Hirayama (1985), the reported temperature ranges from 4000 to 20 000 K. In addition, he suggested that the representative temperature can be set to 7000 K if only one value has to be chosen. Indeed, most researchers found that the temperatures are lower than 10 000 K (Stellmacher, Wiehr, and Dammasch, 2003; Mein and Mein, 1991; Landman, Edberg, and Laney, 1977; Li *et al.*, 1998; Hirayama, 1971). Some researchers, however, obtained temperatures in the higher range of 10 000 – 20 000 K (Ellison, 1952; Jefferies, 1955, 1956). We obtained not only temperatures as low as 4000 K, but also those as high as 20 000 K, including both of the ranges above.

We expect that the temperature might vary along the line of sight. It is well known that a prominence consists of several fine-scale structures (threads). That is, each of the threads has its own temperature, and the temperature we obtained might be the result of the average of several temperatures. For instance, in the multi-thread model of Gunár *et al.* (2008), which fits the UV observations quite well, even identical threads do produce T-variations along the line of sight just because of the random positioning of individual threads. If this is the case, we expect that the intrinsic temperature variation on the plane of the sky may be higher than the variation we observed.

Is the measured peak temperature of 20 000 K real? One might suspect that this is an artifact resulting from large Doppler widths that in fact should be attributed to the superposition of multi-velocity components along the line of sight. Assuming that a resulting high temperature of, for instance, 20 000 K is due to two velocity threads with different line-of-sight velocities, but with the same temperature of 8000 K and the same nonthermal velocity of 6 $\mathrm{km\,s^{-1}}$, what would the relative velocity between the two threads be? From the two-component fitting of the Hα line, the relative line-of-sight velocity between the two components is about 25 $\mathrm{km\,s^{-1}}$. This velocity corresponds to a Doppler shift of 0.7 Å in the Ca II line, which should manifest itself in the two separated components of the spectral line. We observed that the profile of the Ca II line, however, shows only one component with moderate width. This inconsistency indicates that the hypothesis above is wrong, and strongly suggests that the measured high temperatures are real.

We plan to investigate the fine scale structures and temporal changes in the temperature in the prominences in more detail. Higher spatial resolution observations will be available when the next generation of adaptive optics is introduced to the NST. Based on time sequences of high-cadence observations, we will be able to examine the temporal variation in the temperatures of the prominences.

Acknowledgements This work was supported by the National Research Foundation of Korea (KRF-2008-220-C00022) and by the Development of Korean Space Weather Center, a project of KASI.

References

Chae, J., Park, H.-M., Ahn, K., Yang, H., Park, Y.-D., Nah, J., Jang, B.H., Cho, K.-S., Cao, W., Goode, P.R.: 2012, *Solar Phys.* doi:10.1007/s11207-012-0147-x.
Ellison, M.A.: 1952, *Mon. Not. Roy. Astron. Soc.* **112**, 679.
Gallegos, H.G., Machado, M.E.: 1973, *Solar Phys.* **31**, 427.
Gunár, S., Heinzel, P., Anzer, U., Schmieder, B.: 2008, *Astron. Astrophys.* **490**, 307.
Gunár, S., Mein, P., Schmieder, B., Heinzel, P., Mein, N.: 2012, *Astron. Astrophys.* **543**, A93.
Hirayama, T.: 1963, *Publ. Astron. Soc. Japan* **15**, 122.
Hirayama, T.: 1971, *Solar Phys.* **17**, 50.
Hirayama, T.: 1985, *Solar Phys.* **100**, 415.
Jefferies, J.T.: 1955, *Mon. Not. Roy. Astron. Soc.* **115**, 617.
Jefferies, J.T.: 1956, *Mon. Not. Roy. Astron. Soc.* **116**, 629.
Jefferies, J.T., Orrall, F.Q.: 1962, *Astrophys. J.* **135**, 109.
Labrosse, N., Heinzel, P., Vial, J.-C., Kucera, T., Parenti, S., Gunár, S., Schmieder, B., Kilper, G.: 2010, *Space Sci. Rev.* **151**, 243.
Landman, D.A., Edberg, S.J., Laney, C.D.: 1977, *Astrophys. J.* **218**, 888.
Li, K., Schmieder, B., Malherbe, J.-M., Roudier, T., Wiik, J.-E.: 1998, *Solar Phys.* **183**, 323.
Mein, P., Mein, N.: 1991, *Solar Phys.* **136**, 317.
Mouradian, Z., Leroy, J.L.: 1977, *Solar Phys.* **51**, 103.
Nikolsky, G.M., Gulyaev, R.A., Nikolskaya, K.I.: 1971, *Solar Phys.* **21**, 332.
Stellmacher, G., Wiehr, E., Dammasch, I.E.: 2003, *Solar Phys.* **217**, 133.
Tandberg-Hanssen, E. 1995, *The Nature of Solar Prominences*, Kluwer Academic Publishers, Dordrecht, 199.
Zhang, Q.Z., Hu, J., Fang, C., Livingston, W.C.: 1987, *Solar Phys.* **114**, 245.

DOI 10.1007/978-3-319-12123-9_8
Reprinted from *Solar Physics* Journal, DOI 10.1007/s11207-014-0559-x

Active Region Coronal Rain Event Observed by the Fast Imaging Solar Spectrograph on the NST

Kwangsu Ahn · Jongchul Chae · Kyung-Suk Cho · Donguk Song · Heesu Yang · Philip R. Goode · Wenda Cao · Hyungmin Park · Jakyung Nah · Bi-Ho Jang · Young-Deuk Park

Received: 30 June 2012 / Accepted: 3 June 2014 / Published online: 3 July 2014
© Springer Science+Business Media Dordrecht 2014

Abstract The Fast Imaging Solar Spectrograph (FISS) is being operated on the New Solar Telescope of the Big Bear Solar Observatory. It simultaneously records spectra of Hα and Ca II 8542 Å lines, and this dual-spectra measurement provides an estimate of the temperature and nonthermal speed components. We observed a loop structure in AR 11305 using the FISS, SDO/AIA, and STEREO/EUVI in 304 Å, and found plasma material falling along the loop from a coronal height into the umbra of a sunspot, which accelerated up to 80 km s^{-1}. We also observed C2 and C7 flare events near the loop. The temperature of the downflows was in the range of 10 000 – 33 000 K, increasing toward the umbra. The temperature of the flow varied with time, and the temperature near the footpoint rose immediately after the C7 flare, but the temperature toward the umbra remained the same. There seemed to be a temporal correlation between the amount of downflow material and the observed C-class flares. The downflows decreased gradually soon after the flares and then increased after a few hours. These high-speed red-shift events occurred continuously during the observations. The flows observed on-disk in Hα and Ca II 8542 Å appeared as fragmented, fuzzy condensed material falling from the coronal heights when seen off-limb with STEREO/EUVI at 304 Å. Based on these observations, we propose that these flows were an on-disk signature of coronal rain.

Initial Results from FISS
Topical Issue Coordinator: Jongchul Chae

Electronic supplementary material The online version of this article (doi:10.1007/s11207-014-0559-x) contains supplementary material, which is available to authorized users.

K. Ahn (✉) · P.R. Goode · W. Cao
Big Bear Solar Observatory, 40386 North Shore Lane, Big Bear City, CA 92314, USA
e-mail: ksahn@bbso.njit.edu

J. Chae · D. Song · H. Yang · H. Park
Department of Physics and Astronomy, Seoul National University, Seoul, Korea

K.-S. Cho · J. Nah · B.-H. Jang · Y.-D. Park
Korea Astronomy and Space Science Institute, Daejeon, Korea

Keywords Active regions · Chromosphere · Coronal rain · Spectroscopic observations

1. Introduction

There exist several types of active region downflows observed in chromospheric lines, such as inverse Evershed flow and coronal rain. Downflows in active regions and prominences are fairly common (de Groof *et al.*, 2005; Foukal, 1978), and supersonic flows near the sunspot boundary are often seen as well (del Toro Iniesta, Bellot Rubio, and Collados, 2001; Louis *et al.*, 2011). The Evershed flow is observed in the penumbra and in the vicinity of a sunspot. In the photosphere, the Evershed flow is an outward flow (Georgakilas *et al.*, 2003; Ichimoto *et al.*, 2007). In the chromosphere, on the other hand, the direction reverses and the flow is toward the umbra, which is called the inverse Evershed flow (Nicolas and Kjeldseth-Moe, 1981). The Evershed flow is generally along radial threads from the center of a sunspot, and the flows are mostly evenly distributed over the azimuthal direction. Their typical speed is about 2 $\mathrm{km\,s^{-1}}$ in the photosphere, with a line-of-sight speed no higher than 6 $\mathrm{km\,s^{-1}}$ in the chromosphere; some with exceptionally high speeds of up to 46 $\mathrm{km\,s^{-1}}$ (Beckers, 1962; Dialetis, Mein, and Alissandrakis, 1985; Georgakilas *et al.*, 2003).

Coronal rain is material that falls in the form of descending knots along magnetic-field lines (Kawaguchi, 1970; Habbal, Ronan, and Withbroe, 1985; de Groof *et al.*, 2005). Typically, the speed of coronal rain reaches up to 100 $\mathrm{km\,s^{-1}}$, which is much higher than the Evershed flows. The coronal rain is generally recognized as cold material. Its formation is believed to be due to the thermal instability triggered by footpoint heating, such as flares (Goldsmith, 1971; Leroy, 1972; Müller, Peter, and Hansteen, 2004). Field (1965) suggested analytical criteria for the thermal instability to form loop prominences under various physical conditions, including magnetic fields. From a theoretical analysis, Parker (1953) suggested that the characteristic temperature for condensation would be 10^5 and 30 000 K for the transition region and low chromosphere, respectively. This means that the condensation that forms coronal rain would be fairly common. Foukal (1978) claimed that dissipative reconnection might be the source of plasma supply to coronal loops, so that density might increase in the loop. This can also be one of the conditions of condensation, resulting in a downfall motion.

Although the flows of coronal rain may start from coronal heights, they are clearly seen in spectral lines that represent chromospheric temperatures (*e.g.*, Ca II H). Since the flows are faint and fragmented, they are most easily seen outside the solar limb (Habbal, Ronan, and Withbroe, 1985; Antolin and Rouppe van der Voort, 2012), with few on-disk observations (Antolin and Rouppe van der Voort, 2012; Antolin, Vissers, and Rouppe van der Voort, 2012). When seen off the limb, they are in emission. However, when seen on the disk, they are mostly absorption structures (Antolin and Rouppe van der Voort, 2012). While the properties of on-disk coronal rain do not differ much from those of the off-disk rain, identifying coronal rain on the disk will increase the chances for statistical analysis, as there may be more observable events on the disk.

Recently, a new type of mass flow called flocculent flows was observed in the chromosphere (Vissers and Rouppe van der Voort, 2012), which shares characteristics of both inverse Evershed flow and coronal rain. The fragmented shapes and intermittency are similar to the characteristics of coronal rain, but the flows are more aligned with chromospheric fibrils. The flocculent flows are distinct from the coronal rain since they were only seen in the chromosphere.

Here, we have focused on an active-region event, near the disk center, which showed a strong downfall motion. We have analyzed its physical properties that were clearly identified by the spectrograph of the New Solar Telescope of Big Bear Solar Observatory. Section 2 describes the datasets of interest as well as the methods of deducing physical parameters using the cloud model. Section 3 shows the distribution of physical parameters along the flows. Section 4 explains how our observed physical properties and morphology can be linked to various categories of flows observed so far. Finally, Section 5 addresses possible contributions of our findings to the research of high-speed chromospheric events.

2. Data and Analysis

The Fast Imaging Solar Spectrograph (FISS) is the only spectrograph that is currently operational on the 1.6 m New Solar Telescope (NST; Goode *et al.*, 2003). It covers a spectral range of 4000 – 10 000 Å. Various combinations of multiband imaging are possible with the aid of dual CCD cameras with simultaneous imaging capability. The slit width of the FISS is 32 μm, which corresponds to the spatial sampling of 0.16″. The slit height is 8 mm, covering 40″ of the field of view. Detailed technical specifications of the FISS are described in Chae *et al.* (2013).

The theoretical diffraction limits of the FISS in Hα and Ca II 8542 Å lines are 0.085″ and 0.11″, respectively. The spatial resolutions derived from the observed chromospheric features were 0.6″ and 1.2″. These discrepancies mainly stem from the imperfect performance of adaptive optics and the aberrations of the relay optics to the FISS slit. The aberration problem was later corrected, and the spatial resolution in both lines improved considerably.

Active region AR 11305, which was observed on 29 September 2011, is the target of the present study. The observation duration was from 17:10 to 20:06 UT for the FISS, and 201 scans were taken. Figure 1 shows the active region, and its time sequence is displayed in Movie 1 (electronic supplementary material). This active region was located at (295 °E, 122 °N) close to the disk center. The FISS took 400 frames per scan that covered 40″ (in the slit direction) ×60″ (in the scan direction) of the field of view. Each scan took about 60 s, with a cadence of 66 s. This spectrograph has two cameras that can take two spectral bands simultaneously. We chose the Hα and Ca II 8542 Å lines with exposure times of 30 ms and 60 ms, respectively. The camera for the Hα band is equipped with a back-illuminated CCD, which gives about 90 % of quantum efficiency in Hα. The Hα camera has 512 pixels in the wavelength direction and 512 pixels in the slit direction. The second camera for the Ca II 8542 Å band uses a front-illuminated CCD with a quantum efficiency of 35 %, and has 1004 pixels in the wavelength direction and 1000 pixels in the slit direction. Data readout was made in a 1 by 2 binning mode for the Hα camera and in a 2 by 4 binning mode for the Ca II camera. Therefore, one Hα scan made a unit data cube of 512 × 256 × 400 data points, and Ca II 8542 Å data consisted of 502 × 250 × 400 data points. The spectral coverages of the cameras are from −4.9 Å to +4.9 Å for Hα, from −5.8 Å to +6.7 Å for Ca II. Thus, each spectral data point covers 0.019 Å for Hα and 0.025 Å for Ca II.

Every FISS data cube was dark-subtracted and then flat-fielded. Alignment was performed between Hα datasets and Ca II datasets. Since the Hα camera has slightly more pixels in the slit direction (256) than the Ca II camera (250) after binning, the extra area in Hα was trimmed. For the wavelength calibration, we first computed medians of the spectra along the slit at each slit position and then found pixel numbers with the lowest ADU counts at all the slit positions. Finally, we computed the average of all the pixel numbers to determine the line center position with subpixel accuracy.

The alignment with subpixel accuracy was performed among the 201 time-lapse scan data. For a comparison of the data from the FISS, the SDO/AIA, and the STEREO/EUVI

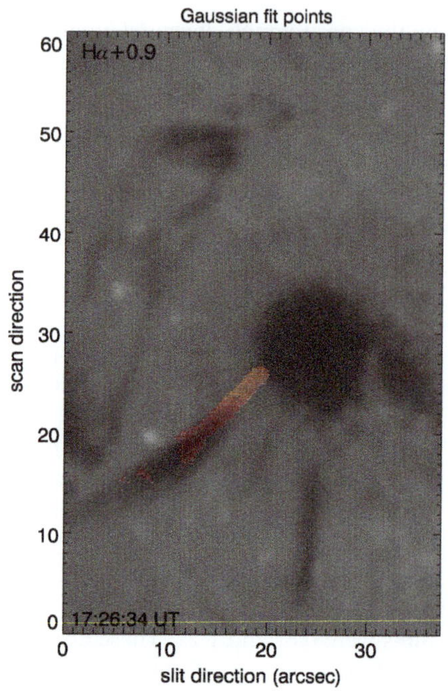

Figure 1 A snapshot of AR 11305 in Hα at a red-wing wavelength of +0.9 Å, marked with the selected pixel points for Gaussian fitting. The points are overdrawn with a color gradient that matches the brightness of the image. These points have features in the red wing of their spectra with a contrast of −0.07 or lower. They also meet successful fitting requirements, so that the fitting profiles are similar to the observed profiles. The field of view is about 40″ (in the slit direction) × 60″ (in the scan direction). The vertical direction in the image does NOT represent the north-south direction on the Sun.

in 304 Å, the orientation of FISS was corrected to be in accordance with the other data, compensating for the rotation rate (360°/day) of the FISS images. There was no image derotating hardware for the FISS; thus we rotated the processed FISS data according to the observation time. No image-restoration technique was used for these slit-scanned data.

All the observations were aided by the NST adaptive optics (AO) system to minimize seeing-induced image motion. This AO system had 76 subapertures, which were not sufficient to ensure diffraction-limited imaging in the blue end of the visible spectrum (Cao *et al.*, 2010). Engineering work is in progress to upgrade this to an AO-308 system, and better spatial resolution for the FISS in bluer wavelengths is expected.

To understand the speed and temperature of our objects, we applied the cloud model (Beckers, 1964) when the objects were isolated from the background, which is typical for our objects. In the cloud model, the contrast profile $C(\lambda)$ is used to identify any absorption or emission feature from the background,

$$C(\lambda) = \frac{I(\lambda)_{\text{out}} - I(\lambda)_{\text{in}}}{I(\lambda)_{\text{in}}} = \left[\frac{S(\lambda)}{I_{\text{c}}}\frac{I_{\text{c}}}{I(\lambda)_{\text{in}}} - 1\right]\left[1 - \mathrm{e}^{-\tau(\lambda)}\right], \tag{1}$$

where $I(\lambda)_{\text{in}}$ is the background intensity, $I(\lambda)_{\text{out}}$ is the intensity of the target object, $S(\lambda)$ is the source function, I_{c} is the continuum intensity, and $\tau(\lambda)$ is the optical thickness. The source function was assumed to be dominated by scattering from background radiation. We used the Hα modeling of Heinzel, Gouttebroze, and Vial (1994) and Heinzel and Anzer (2006), namely $S/I(\lambda_0)_{\text{in}} = 0.5$ (λ_0 means the line center wavelength), and also assumed that this value holds for the Ca II 8542 Å line. This assumes that the source function is nearly constant by complete redistribution (Vardavas and Cram, 1974). The ratios $I_{\text{c}}/I(\lambda)_{\text{in}}$ for both spectral lines were adopted from the atlas of the solar spectrum (Wallace, Hinkle, and Livingston 1993, 1998). The background intensity was chosen from the vicinity of a pixel that did not contain the loop of interest over the entire time range. We calculated this time-averaged background intensity as a function of distance from the penumbra in the loop direction. Different continuum levels between the target spectrum and the reference spectrum were adjusted so that the contrast of the wing side would be close to 0.

Finally, we fit the contrast profile with a Gaussian profile as a function of optical thickness τ,

$$\tau(\lambda) = \tau_0 e^{-(\frac{\lambda-\lambda_D}{\Delta\lambda_D})^2}, \tag{2}$$

where τ_0, λ_D, and $\Delta\lambda_D$ are the peak optical thickness, Doppler shift, and Doppler width of the Gaussian profile.

By assuming that there is a single velocity thread without multiple bulk motions, its spectral line width is mainly due to the Doppler broadening

$$\Delta\lambda_{D(H\alpha)} = \frac{\lambda_{H\alpha}}{c}\sqrt{\frac{2kT_{H\alpha}}{m} + \xi_{H\alpha}^2}, \tag{3}$$

$$\Delta\lambda_{D(Ca)} = \frac{\lambda_{Ca}}{c}\sqrt{\frac{2kT_{Ca}}{m} + \xi_{Ca}^2}, \tag{4}$$

where

$$T_{H\alpha} \simeq T_{Ca}, \tag{5}$$

$$\xi_{H\alpha} \simeq \xi_{Ca}. \tag{6}$$

Here, m is the mass of an element that is involved with the spectral line, and ξ is the non-thermal speed. We assumed that the plasma measured in both spectral bands was identical or asymptotically so. This may not be always true since different spectral lines may represent different formation temperatures.

The simultaneous dual-band spectroscopic capability of the FISS enables distinguishing plasma temperature from nonthermal contribution to the line broadening. This capability is maximized when we observe two lines with different atomic mass. We used the hydrogen and calcium lines. Since the atomic mass of calcium is much higher than hydrogen, the thermal line-broadening component was easily extracted.

We chose the most prominent loop with an absorption feature in the field of view of the FISS. To identify a high-redshift feature in the loop, we cropped a spectral window from the contrast profiles in the red wing of each spectral band (from +0.46 Å to +2.7 Å for Hα, from +0.7 Å to +4.2 Å for Ca II). Then we chose the pixel with the lowest contrast for every spectrum to roughly locate the line center. The line profile to be fitted was narrowed down to ± 100 pixels for Hα and ± 20 pixels for Ca II from the pixel with the lowest contrast. We selected image pixels in the loop by setting the threshold contrast to −0.07 or lower for both lines.

From the selected contrast profiles, we performed Gaussian fitting on the calculated $\tau(\lambda)$ profiles, and we only accepted pixel points when λ_D and $\tau(\lambda_D)$ values derived from the observed $\tau(\lambda)$ profile data and the fitted profiles matched within 0.1 Å and 0.1, respectively. In this manner, we identified about 1000 pixels from a scan that fulfilled our criteria. After fitting with Gaussian profiles, the velocity for each pixel in the loop was derived by choosing the wavelength with the lowest contrast from the fitted profiles. The temperature and the nonthermal components of the pixels were calculated from the Doppler widths derived from the Gaussian fittings.

In order to understand what triggered these high-redshift flows, we have compared our data with SDO/AIA EUV level-1 context images at 171 Å and 304 Å (Lemen *et al.*, 2012). We used the Solar Software (SSW) routines to process data. These images have a scale of 0.6 arcsec/pixel. The 171 Å images are suitable for understanding coronal loop structures of about 0.6 MK, while the 304 Å images show high chromospheric structures (5×10^4 K). In total, 1217 images were chosen in each wavelength with a 36 s cadence from 09:00:09 UT to 21:09:45 UT.

To understand the orientation of the high-redshift flows and their relation to the surge and the flares, we examined the STEREO-B/EUVI images in 304 Å in JPEG format for context analysis. At the time of observation, the STEREO-B spacecraft was located at a trailing location in the Earth's orbit by about 90°. We chose 141 images from the STEREO observation on the same day with 10 min cadence. The image scale of the EUVI data is 1.6 $\mathrm{arcsec\,pixel^{-1}}$. Among the RGB image channels of jpeg images, we only chose the G channel, since this gave the best contrast of the coronal flow events against the dark background.

3. Results

3.1. Morphology

Figure 2 shows multiwavelength structures of AR 11305 observed in Hα and Ca II 8542 Å lines. Images constructed from the line center intensities of the lines display a closed chromospheric loop that connected two sets of pores (marked A). The apex of the loop is clearly seen in the blue wing of Hα (panel b), while the footpoints are more clearly seen in its red wing (panel e). Based on these findings, this structure seemed to be a rising loop lifting hydrogen material, while the material soon drained into the two loop ends. This closed loop in the Ca II images, on the other hand, showed only the lower part of the loop with an absorption feature in the line center as well as wing images (panels g, h, and i). The low visibility at the far side of the loop from the active region in Ca II implies that this part is optically thinner in Ca II than in Hα. The footpoints of the loop were bright in the line center images of both lines (panels c and h), probably implying material collision due to the downflows.

The image made from the far red wing of Hα (panel e) shows several loops where high-density plasma flowed into the umbra of the sunspot (marked B and C). No counterpart of these flows was observed in the blue wing, which suggests that only downflows occurred in the loops. The image made from the far red wing of the Ca II 8542 Å (panel j) shows only a small portion of this falling material indicated as a faint, dark blob in a square. We focused on loop B for our analysis, since only this gave sufficient contrast in both Hα and Ca II.

3.2. Spectral Properties

The flow patterns are not directly obvious from these images, but the spectral data in Figure 3 clearly show that there existed strong redshift flows along the line of sight. Since this sunspot was located near the disk center, we inferred that these redshift velocity components represented mostly vertical motions on the solar surface. The flows were as fast as 80 $\mathrm{km\,s^{-1}}$, and this velocity was high enough to create a distinct line dip in both Hα and Ca II, as shown in the contrast profiles of Figure 4. After the flows of the loop merged with the umbra, we were unable to identify the terminal speed of the flows because of low contrast.

Since an image taken at a single wavelength position cannot provide a whole coverage of the flows, we infer a high-redshift gradient for the Doppler velocity along the loop. While

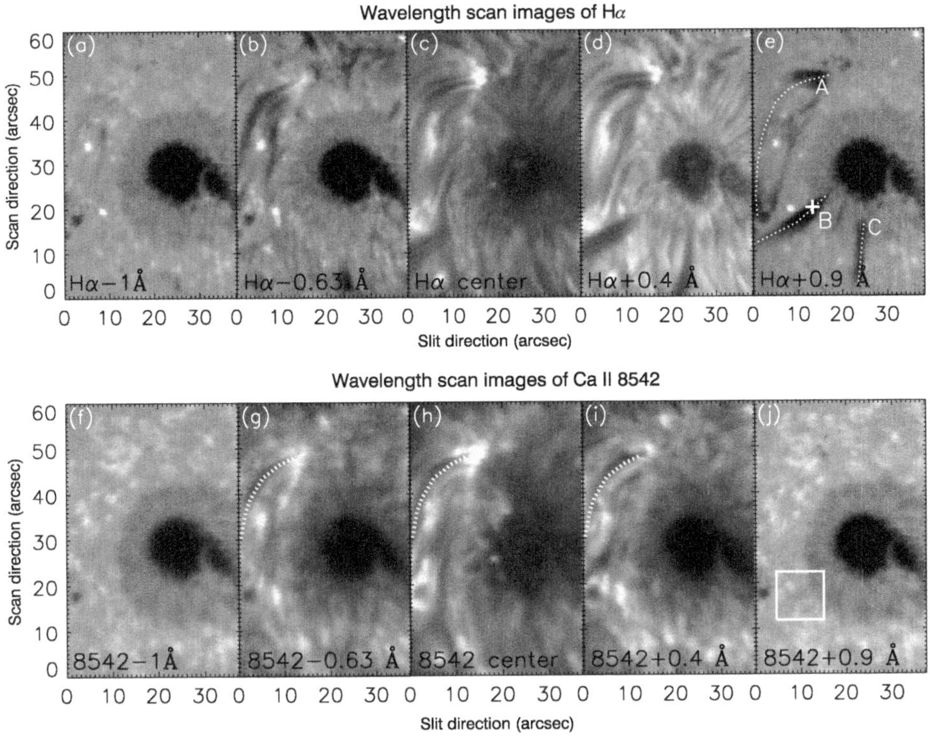

Figure 2 AR 11305 in Hα (upper row) and Ca II 8542 Å (lower row) at different wavelengths. The images were taken at 17:26:34 UT. Each panel has a field of view of 40 arcsec (horizontal) and 60 arcsec (vertical). Features A, B, and C marked in panel (e) denote: A: a closed loop that showed relatively horizontal flows; B and C: apparently open loops that showed only downflow motions in Hα. Square: the same loop as B, but visible in Ca II 8542 Å as a faint and dark structure in the middle of the square.

the Hα images show elongated loop structures such as B of Figure 2, the Ca II slices only show an absorption feature that covers a small fraction of the loop. If the contrast values in both lines do not change significantly along the loop, we can infer that this difference comes from the different line widths of Hα and Ca II. The Ca II 8542 Å image slices show much smaller line broadening because of the high atomic mass of calcium. Thus, only choosing one wavelength was an inadequate approach to understand the morphology of this loop.

Therefore, in order to enhance the visibility of the loop structure we have integrated over the wavelength along the spectral line. From the contrast profiles, we identified the absorption profile of the redshift flows. We therefore integrated the contrast profiles along the spectral ranges that were associated with these flows.

Figure 5 shows the maps of wavelength-integrated contrast constructed to enhance the loop structure. The loop structure as seen in both lines represents a much wider velocity coverage. Owing to the low contrast in the features in Ca II 8542 Å, only one loop was clearly visible co-spatially in both lines. This co-spatial feature provides reliability in the determination of temperature and nonthermal speed.

The relatively low contrast of this feature in Ca II 8542 Å can be attributed to the lower optical thickness of the Ca II plasma, which can be seen in most of FISS observations with chromospheric targets. In fact, the low-contrast feature in the Ca II 8542 Å band has been expected from cloud modeling (Ahn *et al.*, 2008). It was also found in filament observations

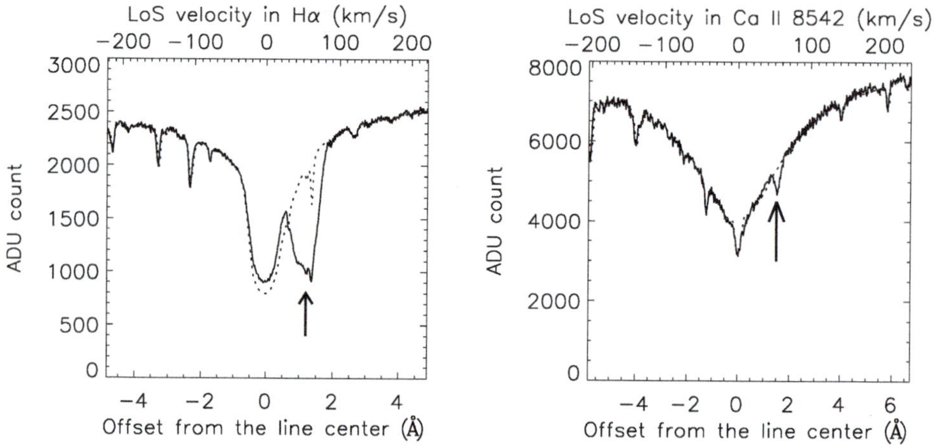

Figure 3 FISS spectral line profiles at a position in loop B (marked with '+' sign) of Figure 2 in both wavelengths (solid), drawn with reference profiles (dotted). The reference profiles were chosen from the background in the vicinity of the point that lay outside of the loop. The components with high Doppler shifts are indicated by arrows.

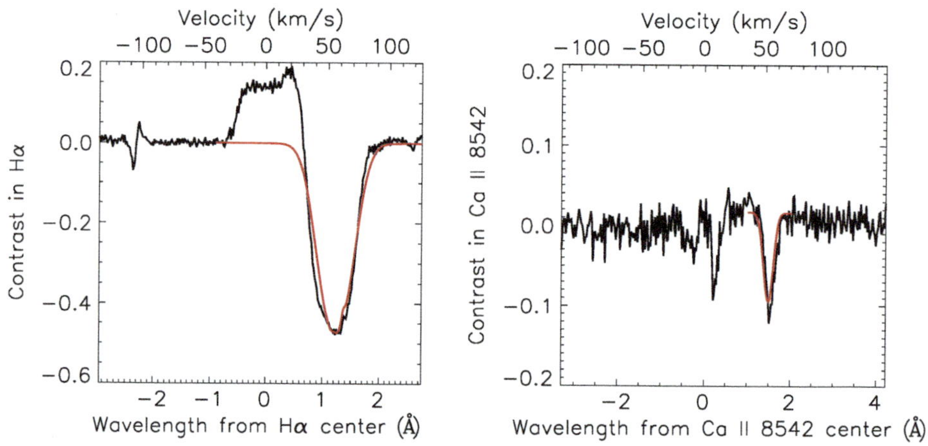

Figure 4 Contrast profiles of the high-redshift feature from the line profiles in Figure 3, with high-redshift features indicated in both wavelengths with fitted spectral profiles (red curve).

in which the Ca II absorption feature mostly filled the lower part of filaments (Heinzel and Anzer, 2006; Tziotziou *et al.*, 2001).

To show the change in the line width and Doppler-shift patterns, we selected 11 sample points along the length of the loop of interest (see Figure 6). The selected points are marked with '+' signs with index numbers next to them. Because of the low optical thickness of Ca II 8542 Å in the outer part of the loop, the fitting did not work reliably at the points from 0 to 5 and we were unable to perform accurate fittings in these points. Nevertheless, an increase in the line width as we approach the umbra in Hα is seen (lower left panel of Figure 6), while the Ca II 8542 Å line did not show a noticeable line width change. Owing to the low atomic mass of hydrogen, the line width of Hα is much more sensitive to temperature than Ca II 8542 Å, while the nonthermal component may affect both spectra

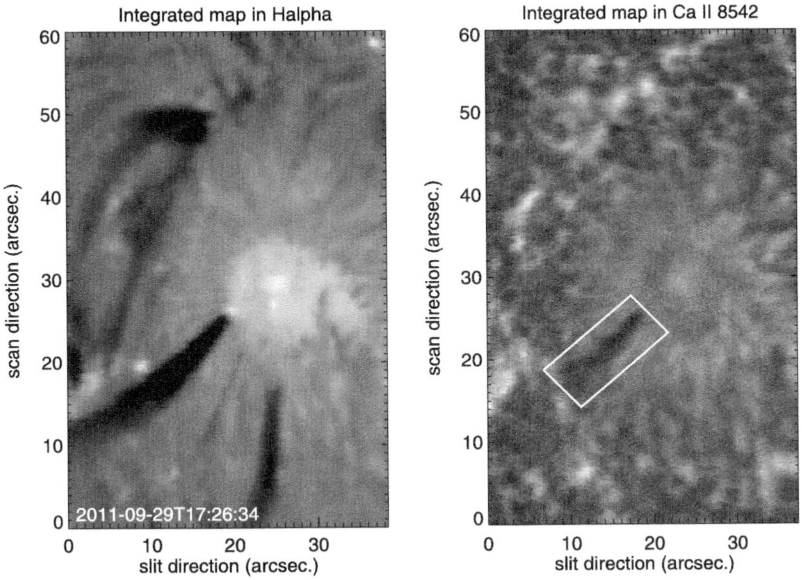

Figure 5 Loops with falling plasma in AR 11305 (shown as dark elongated loops), in the red-wing part of Hα (left) and in the Ca II 8542 Å line (right). Both images were taken at 17:26:34 UT. The plasma in the loop seen in Ca II is shown inside a rectangular box. The integrated map in Hα is a result of wavelength integration from +0.28 Å to +1.80 Å with 81 wavelength positions. The map in Ca II resulted from an integration from +0.96 Å to +2.20 Å with 51 wavelength positions.

in the same weight. Thus, the fact that only Hα showed significant increase in line width means that thermal broadening was the dominant source for the increase in the line width. Meanwhile, both spectral lines showed an increase in the Doppler shift.

3.3. Temporal Evolution

Figure 7 shows a snapshot of the SDO/AIA images in the EUV wavelengths, and their temporal evolution (see Movie 2) shows that there were flaring activities, a C2 flare at 10:22:21 UT and another C7 flare at 18:03:09 UT. The C7 flare was covered by the FISS observation. There were several bright surge events from 16:40 to 18:00 UT, which is clearly visible in the bottom panel of Figure 7. The surges were bright at the rising phase, but then decreased in brightness so that their contrast against the background became low. The loop with observed redshift events in the FISS data was identified as an elongated absorption loop that arched above the bright active region in 304 Å images (shown in box D with a dotted line). This downflow in 304 Å was mostly visible for the entire time sequence of the 304 Å images. We were unable to identify the outermost location of this loop in either Hα and He 304 Å. The field of view of the FISS was not sufficient to cover the whole loop. The loop in He 304 Å was extended to a dark background, which gave little contrast of the loop against the background.

The Hα time-lapse images in Figure 8 (and Movie 1) show that the amount of plasma in the high-redshift loop decreased immediately after the flare. The amount of the high-redshift flows along the loop decreased about 1 h after the flare (panels (5) and (6) of Figure 8), the amount of downflow material that did not apparently reach the umbra considerably increased in another bundle of loops in the north-west part of the active region and lasted for another 1 – 2 h (panels (5) and (6)) during the FISS observation.

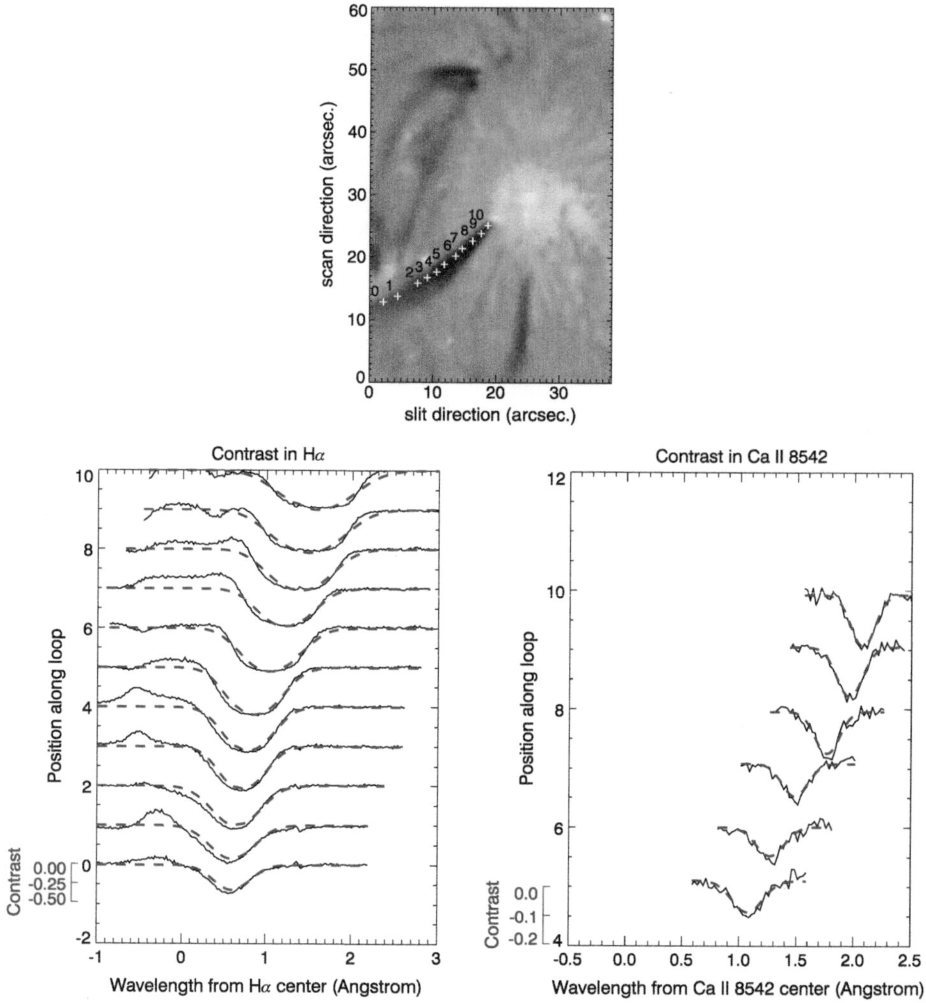

Figure 6 Top: numbered sample pixel positions in an integrated Hα image taken at 17:26:34 UT. The integrated map in Hα is a result of wavelength integration from +0.28 Å to +1.80 Å with 81 wavelength positions. Bottom: contrast profiles (solid lines) calculated at the numbered positions in Hα (left) and Ca II 8542 Å (right) lines, overplotted with Gaussian-fitted profiles (dashed). Positive integer offsets corresponding to the numbers marked in the top panel were applied in the y-direction, so that all the profiles may be visible separately. The wing of each contrast profile should be close to the marked number in the y-direction. For the Ca II 8542 Å case, only the positions from 6 to 10 were drawn, since the Gaussian fitting worked only at these positions. Contrast scales are displayed on the left axis of each panel.

Figure 9 shows a selected sequence of cropped images of STEREO-B/EUVI in 304 Å, and the time evolution can be seen in Movie 3. Panel (1) shows the fragmented blobs in the loop before the C7 flare event. During most of the time sequence, however, it was difficult to identify individual blobs because they appeared as a connected flow. This may be partly because of the insufficient spatial resolution of the instrument. Panel (2) shows the onset of a surge, which is also seen in the SDO data (see Figure 7). Relatively dark and low-lying elongated material in the southwest direction corresponds to the loop that involved the high-

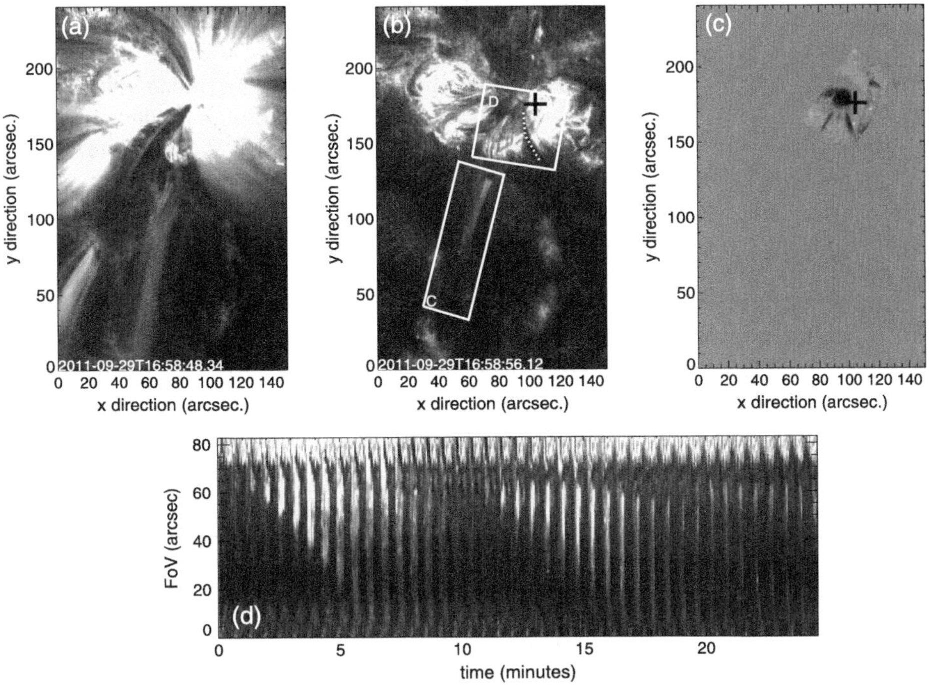

Figure 7 A snapshot of the SDO/AIA data. Panel (a): 171 Å, panel (b): 304 Å, panel (c): Hα for a field-of-view comparison. All the images were taken nearly at the same time. A bright surge from the AR in rectangular box C is shown in panel (b), while multiple thread-like absorption features are seen in box D. (d) Time-slice image of the surge in box C of panel (b), with $t = 0$ at 16:42:09 UT. An elongated absorption feature with the downfall motion seen in Hα (panel (c)), extending to the 5 o'clock direction in the display, is marked with a dotted line in box D. The location of a C7 flare is marked as a black cross in panels (b) and (c).

redshift flows. Panel (3) shows the maximum growth of the surge, marked with a dotted line. From this perspective, the 304 Å loop appears to lie below the surge, while there is no guarantee that both events were co-spatial. Panel (4) corresponds to the maximum phase of the C7 flare. The flare partially illuminated the surge, but the brightening along the solar surface in the southern direction was much longer. After the flare event, the 304 Å loop became much fainter (see panels 4 and 5). After 4–5 h, the material in the 304 Å loop began to reappear along the loop and again flowed toward the active region.

3.4. Velocity

Figure 10 depicts the Doppler velocity profile along the sampled ≈ 1000 points in Figure 1. While the overall trend is linear along the loop distance, the velocity increases much more steeply along the loop down to 6 arcsec from the umbral side, which is followed by a flatter increase. When the Doppler component reached the line center that contained the background profiles, it became difficult to define the lowest values at these points. Such points were mostly excluded from the profiles, but there are still some points that are out of the linear trend. The velocity near the far end of the loop was about 25 $\mathrm{km\,s^{-1}}$, while it increased up to about 80 $\mathrm{km\,s^{-1}}$ as the flow reached the umbra. The Doppler velocities measured both in Hα and Ca II 8542 Å matched well.

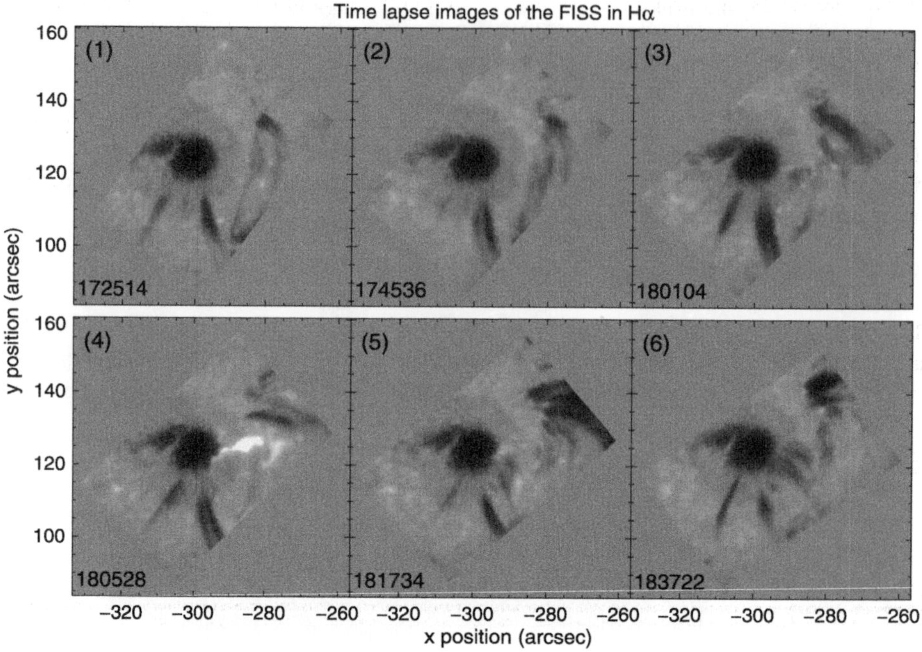

Figure 8 Time-lapse images of the Hα red wing (+0.9 Å) around a C7 flare event. The time of observation (hhmmss) is indicated in each panel. The elongated white patch in panel (4) corresponds to the C7 flare event.

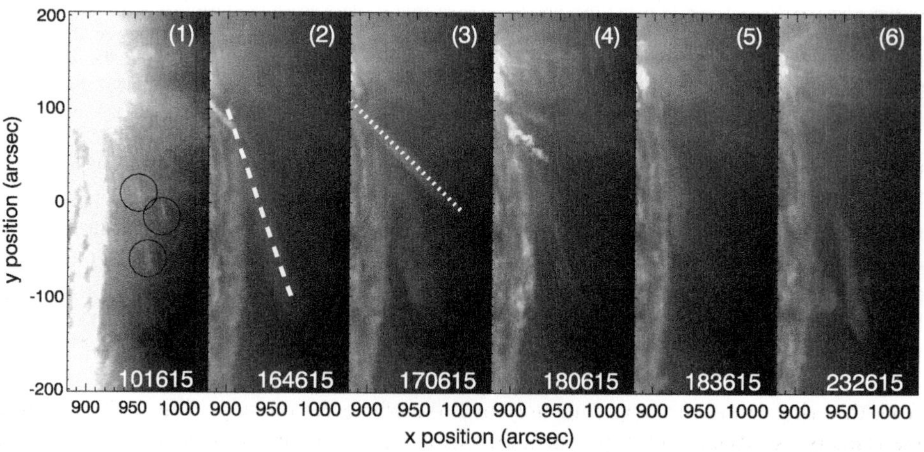

Figure 9 Time-lapse images of STEREO-B/EUVI at 304 Å. The time of observation is shown at the bottom of each panel. The gray scale of panel (1) was saturated at a medium intensity to enhance the contrast of the fragmented blob in the loop. The individually identified blobs are circled. Panel (2) shows the inclined loop with a dashed line that contains the redshift flows in the FISS observation. Panel (3) shows an elongated surge marked with a dotted line.

Figure 11 shows a time-distance plot of the flows in the loop from STEREO/EUVI in 304 Å. Even though it looked faint, there were inclined features that indicated flows toward the active region. The most prominent flow is marked with a solid line. The slope of this line

Figure 10 Velocity distribution as a function of distance from the active region. These velocities were separately measured from the contrast profiles in Hα (black) and Ca II 8542 Å (green) from the marked positions in Figure 1. We performed separate Gaussian fitting for Hα and Ca II, and only the points with converged Gaussian fitting are plotted. The gap between 14 and 16 arcsec in Ca II means that the fitting failed in this range in Ca II while it was satisfactory in Hα.

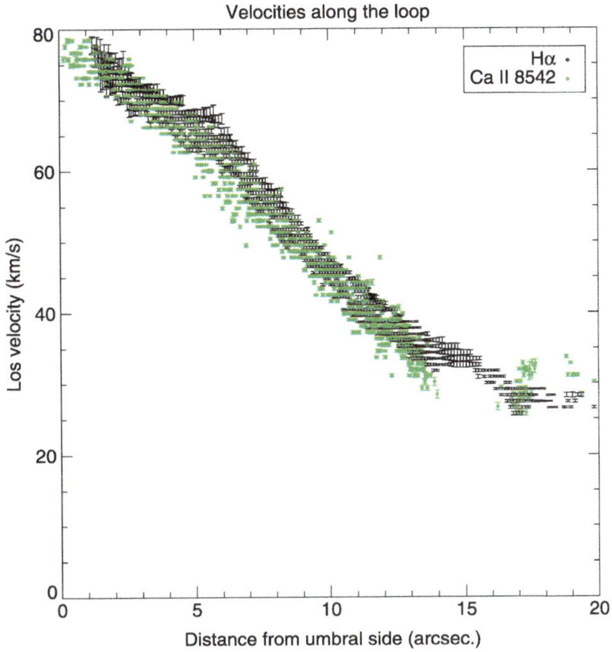

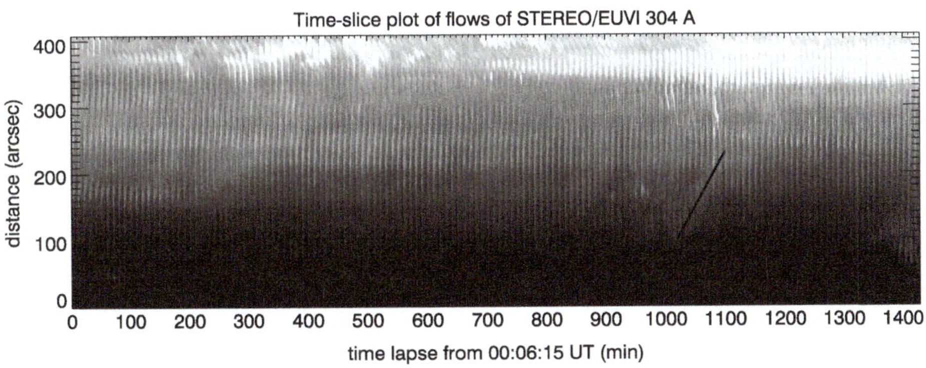

Figure 11 Time-slice plot of flows observed by STEREO/EUVI in 304 Å, along the direction marked as a dashed line in Figure 9 on panel (2). The black line in the figure represents the most prominent flows detected in this time-lapse data.

yielded a speed along the loop of about 20 km s^{-1}. This loop was nearly parallel to the solar surface, making an angle of about 20°. Considering that the loop lay approximately in the north-south direction, we estimate that its lowest vertical speed was close to $20 \tan 20° \approx 7$ km s^{-1}, with a lowest horizontal speed of $20 \cos 20° \approx 19$ km s^{-1}. The location of the visible velocity component in Figure 11 was about 150 arcsec away from the active region in the horizontal direction. The vertical speed was much lower than the lowest line-of-sight speed measured by the FISS (about 25 km s^{-1} at 20 arcsec away from the active region). We estimate that the velocity of the flow has slowly accelerated from about 7 km s^{-1} to about 25 km s^{-1} as it traveled by about 130 arcsec. The STEREO images covered the upper part

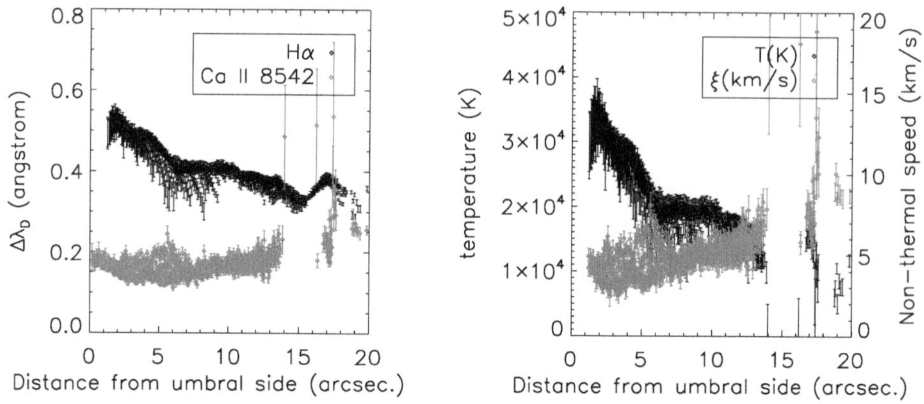

Figure 12 Left: measured Doppler widths from the sampled positions in Figure 1, in Hα (black dot) and Ca II 8542 Å (gray diamond). Right: calculated temperatures (black dot) and nonthermal components (gray diamond). Only pixel points with converged Gaussian fitting both in Hα and Ca II are included.

of the loop that extended more than 100 arcsec along the solar surface, while the FISS data only covered about 20 arcsec from the active region.

3.5. Thermodynamics

Figure 12 shows the temperature distribution of the downflows along the loop; we deduced the temperature of the material in the loop to be in the range of 10 000 – 33 000 K. Closer to the umbra, the temperature increases up to $33\,000 \pm 5000$ K, which is above the hydrogen-ionizing temperature. Meanwhile, the nonthermal speed was about 8 ± 1 km s^{-1} at the far side of the loop, and then it decreased to 3 ± 1 km s^{-1}. There is an empty window between the distances of 14 and 16 arcsec. This is mainly due to failed Gaussian fitting in the points within this range, since the wing of the Ca II line profiles was contaminated by the superposition of the line-center profile that had nonzero contrast values (see Figure 4). For the same reason, the temperature and nonthermal speed beyond 16 arcsec show large scatter. Some of them are probably due to fitting errors.

Figure 13 shows the time-lapse images of the temperature map of the downflows in the loop from 17:26:34 UT to 18:42:52 UT, drawn on Hα red-wing images as background. The loop of interest is visible in most images of this sequence. The temperature of the downflows in the loop was around 10 000 K at the far side of the loop from the active region. We can clearly see the increase in temperature along the loop toward the active region throughout the sequence. Panels (4) and (5) show that the loop was split into two loops at the far side from the umbra. One of the divided loops toward the left in the slit direction showed a higher temperature. The upper portion of this loop seemed to have two temperature peaks along the loop, one in the middle of the measured loop, the other at the near-umbra side. The other of the two divided loops was relatively cooler with a peak temperature of about 28 000 K.

The far side of the loop did not seem to change its temperature throughout the observation. However, the peak temperature at the near side of the loop toward the umbra varied often. The footpoint of the loop showed a high-temperature signature (panels (1), (3), (5), and (6)), and the peak temperature fluctuated between 28 000 K and 35 000 K. The highest temperature corresponded to the time of panel (6), five minutes after the C7 flare event. This suggests that the footpoint heating was triggered by the flare event. One interesting aspect

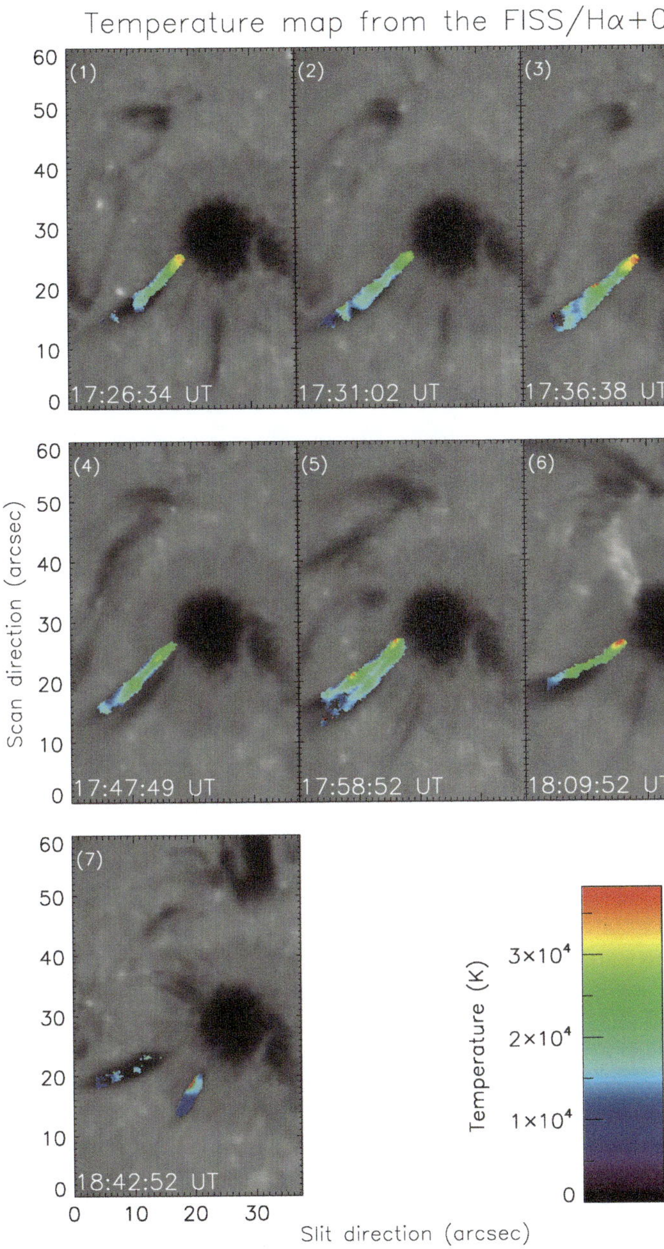

Figure 13 The time-lapse images of temperature distribution in the downflows in the loop. The background images are taken from the red wing of Hα at +0.9 Å. The color scale for the temperature is shown at the right side of panel (7).

to be noted in this panel is that the highest temperature was seen in the leftward divided loop in the map. The rightward loop kept its lower temperature of about 27 000 K even after the flare. Many factors are expected to contribute to heating the footpoints of the loops, but

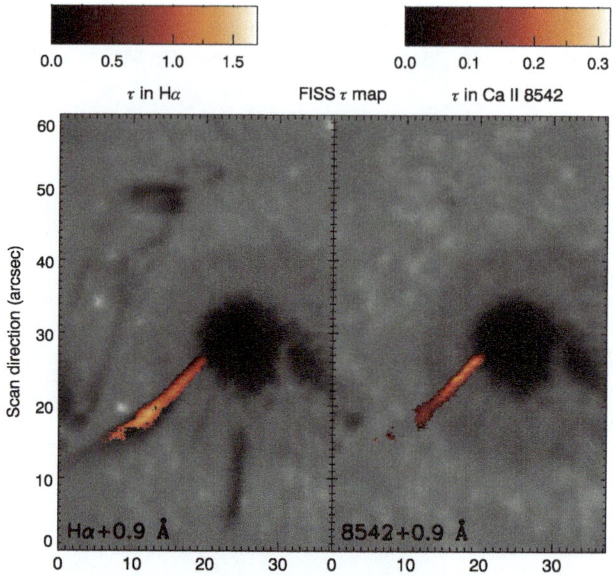

Figure 14 Spatial distribution of peak optical thickness (τ_0) along the loop in both wavelengths of the FISS. Background images are made from the red-wing intensities of the two wavelength bands.

the temporal correlation with the flare event suggests that there were at least two loops with different temperature structures, and the flare heated only the leftward loop.

Figure 14 depicts the spatial distribution of the peak optical thickness (τ_0) derived from the cloud model. The Hα line showed an optical thickness about five times as high as the Ca II 8542 Å line. The optical thickness of Ca II was higher in the lower part of the loop, while the optical thickness of Hα increased toward the top of the loop.

4. Discussion

Our observed loop showed redshift flows all the way up to the corona. The loop extended from the coronal height and might be linked to the umbral region of the sunspot as its footpoint. The loop showed a redshift of about 80 km s^{-1}. Its temperature ranged between 10 000 and 33 000 K, which is within the chromospheric temperature range. It seems to be higher than typically observed temperatures of Hα and Ca II 8542, which is lower than 10 000 K (Mein *et al.*, 2000; Tziotziou *et al.*, 2001; Antolin and Rouppe van der Voort, 2012), although the Hα temperature was extended to up to 5×10^5 K in a previous observation (Antolin and Rouppe van der Voort, 2012). The optical thickness of Hα reached up to 1.5, whereas for Ca II 8542 Å it was 0.3.

Based on the present observational results, we propose that the flows in the loop were the signature of the coronal rain. This is mainly because we were only able to see recurring (and occasionally fragmented) redshift motion in Hα, Ca II 8542 Å, and He 304 Å observations. Considering that this active region was located close to the disk center, we infer that the redshift motions seen in the FISS spectra corresponded to downflows. Unlike ordinary coronal rain observations, the flows looked like continuous flows rather than fragmented blobs in most cases. However, we interpret this as a combination of numerous blobs that remained unresolved by the spatial resolution of the FISS and other instruments, because we were occasionally able to identify fragmented blobs from STEREO observations when the blobs were sparse. Such bursts of numerous condensations have been described as 'coronal showers' by Antolin and Rouppe van der Voort (2012). Still, there is a possibility that these flows might be a part of siphon flows inside a closed loop, since we did not clearly notice the entire loop structure in the observation. However, the falling material began to appear after passing the apex of the loop, which strongly suggests that condensation was in progress.

The question remains whether the Doppler line width really came from a single component that had a unique Doppler velocity. If multiple velocity threads with different Doppler velocities overlapped along the line of sight, then a wider line width, exceeding the thermal width, would have been observed. Since the thermal line width of Hα is large, such multi-velocity components might overlap and might be difficult to identify individually. In fact, from the time-lapse image sequence of Hα (Movie 1), sometimes there seemed to be at least two loops that were nearly at the same position. While these two loops were separated at the far side of the sunspot, they apparently merged into a single structure near the umbra. If these two loops had different Doppler velocities, then these two different loops along the same line of sight could be the reason for the increase in the line width near the umbra. Conversely, the Ca II line has a much smaller line width, and we could easily have identified multivelocity or multitemperature threads from its spectrum, as shown in Figure 13, had they existed. Nevertheless, the possibility of feature smearing remains because of the finite spatial resolution.

Throughout the STEREO/EUVI observations at 304 Å that lasted for 24 h, we saw recurring downflows in the loop. The question therefore arises how this persistent mass supply was possible. We detected many small-scale flare events throughout our observation, and the amount of visible material was affected by the C-class flares. Immediately after the C-flare events, the amount of falling mass decreased, and several hours later, we began to see the initiation of condensation. We consider the decrease in plasma after the flare events as heating of the plasma above the formation temperature of He II. Such footpoint heating might simply be a mechanism of mass supply to the coronal loop by chromospheric evaporation (Goldsmith, 1971; Antiochos *et al.*, 1999; Antolin, Shibata, and Vissers, 2010), or these flare events might trigger dissipative reconnection (Foukal, 1978), so that mass from different loops moved across the magnetic-field lines.

From the FISS images, the plasma in the loop was faint at the far side of the active region in Hα and Ca II 8542 Å, but was clearly seen near the active region. On the other hand, the absorption feature seemed to be much more extended toward the outer part of the active region in the He 304 Å images. The formation temperature of the He 304 Å line is known to be higher than that of Hα and Ca II 8542 Å (Wang *et al.*, 1998), and this might explain the different views in different data. The increasing visibility in Hα and Ca II 8542 Å can be explained by the catastrophic cooling process that is triggered by thermal instability (Parker, 1953; Field, 1965; Schrijver, 2001). As the heated plasma rapidly cools to the chromospheric temperature, recombination takes place. Eventually, this process is detected as absorption in Hα and other wavelengths. The frequency of the coronal rain production decreased immediately after the flare event; however, it increased again after a few hours. We propose that the condensation process was suppressed after the loop was heated by flares and restarted as it cooled after a few hours.

We realize that STEREO observed the top part of the loop, while FISS data only covered the footpoint side of about 20 arcsec length. Thus, it was difficult to find an overlapping area between the two datasets to construct a three-dimensional picture of the loop. However, it gave a consistency that the vertical speed of the flow observed from STEREO was slower than the lowest speed derived from FISS Hα, which implies that the acceleration was already initiated before the flow reached the field of view of the FISS.

Interestingly, the temperature of the downflows along the loop structure increased as the flows moved toward the umbra. The most obvious energy source for the temperature rise is gravitational energy. It is known that the falling coronal rain has an acceleration below gravity or even deceleration (Antolin, Vissers, and Rouppe van der Voort, 2012; Müller, Hansteen, and Peter, 2003; Müller, Peter, and Hansteen, 2004), indicating the existence of resistance.

Gravitational energy has been reported to be a sufficient source for loop heating (Hasan and Schuessler, 1985; Narain and Ulmschneider, 1990). It is also possible to estimate the amount of energy released during the downfall motion. Because the loop was located near the disk center, we regarded the Doppler speed as the vertical component of the flows. Though we cannot assume that the flows were in free fall, we can calculate the minimum height of the loop from the free-fall assumption. From the terminal velocity of 80 $\mathrm{km\,s^{-1}}$, we estimate the minimum loop height as 11 600 km. Some observations reported that statistical average acceleration of coronal rain events is about $1/3$ of solar surface gravity (Antolin and Rouppe van der Voort, 2012; Schrijver, 2001). In this case, the real height of the loop would be three times higher, which is about 34 800 km for our case. This height is well above the chromospheric layer, and this confirms that what we observed was coronal rain.

Considering the temperature increase in a microscopic picture, we can estimate the required gravitational energy difference to achieve both the temperature (from 10 000 K to 33 000 K) and the speed increase (from 0 to 80 $\mathrm{km\,s^{-1}}$) from the following equation:

$$\mu g\,\mathrm{d}h \geq \frac{3}{2} Nk\,\mathrm{d}T + \frac{1}{2}\mu\,\mathrm{d}(v^2), \tag{7}$$

where $\mu = m_\mathrm{H}N_\mathrm{H} + m_\mathrm{p}N_\mathrm{p} \approx m_\mathrm{p}N_\mathrm{p}$ and $N = N_\mathrm{H} + N_\mathrm{e} + N_\mathrm{p} \approx N_\mathrm{e} + N_\mathrm{p} = 2N_\mathrm{p}$ if we assume a significant ionization fraction.

The gravitational energy difference required for the temperature increase is 2100 km, and if we add 11 600 km for the velocity increase, we derive the total height difference required as $\mathrm{d}h = 13\,700$ km. This is far below the estimated real height of the loop (about 34 800 km), which implies that the gravitational energy is sufficient to play a role as a heating energy source. If we assume that this loop was part of an arc shape, we can speculate that the increase in effective gravity along the loop probably plays a role. The linear-like velocity trend along the loop also supports this assumption.

After the catastrophic cooling in the higher part of the loop, the temperature seemed to increase again toward the loop footpoint. Although the gravitational energy might be a source of such an increase in the lower part of the loop, the temperature rise did not seem to be large enough to trigger another catastrophic cooling process.

5. Conclusions

We observed a coronal-rain event that was accompanied with small-scale flares in an active region. The temperature of the coronal rain flows increased to exceed the ionization temperatures of the Hα and Ca II 8542 Å lines as the flows reached the footpoint of the loop, but these flows were still visible in these lines. We also found that the flare events, in small-scale or micro-scale, were dominant triggers to change the amount of coronal rain material. There may be many different structures with chromospheric temperatures at coronal heights, and it is sometimes very difficult to distinguish these structures. Determining velocity, temperature, and their morphology along the structures might be good indicators to facilitate this categorization. Multichannel observations using different spectral lines and wavelength bands were helpful in this analysis. We emphasize that the multiviewpoint observations using STEREO data were very crucial to disambiguate superposed structures, particularly when they were very close to each other. We plan to perform more observations using the same technique on more active regions. It will be particularly helpful if the new observations have higher spatial resolution. We will definitely be able to achieve higher spatial resolution with the aid of new AO-308 system as well as improved FISS relay optics with less aberration on the NST. This higher resolution will provide a higher sensitivity for detecting multiple loops in a small area, so that we will be able to explore an even wider range of physical properties (such as temperature) of the flows in loops. Since there have not been many studies regarding the physical properties of the coronal rain seen on the disk, it will be interesting to see the variation of these in relation to active regions of different types.

Acknowledgements I would like to thank the referee for many critical comments, which improved this manuscript considerably. This research was supported by the National Research Foundation of Korea (355-2012-1-2012R1A6A3A03039815 and 2012R1A2A1A03670387). K. Ahn and W. Cao gratefully acknowledge the support of NSF-CAREER through ATM-0847126.

References

Ahn, K., Chae, J., Park, H.-M., Nah, J., Park, Y.-D., Jang, B.-H., Moon, Y.-J.: 2008, *J. Korean Astron. Soc.* **41**, 31.
Antiochos, S.K., MacNeice, P.J., Spicer, P.J., Klimchuk, J.A.: 1999, *Astrophys. J.* **512**, 985.
Antolin, P., Rouppe van der Voort, L.: 2012, *Astrophys. J.* **745**, 152.
Antolin, P., Shibata, K., Vissers, G.: 2010, *Astrophys. J.* **716**, 154.
Antolin, P., Vissers, G., Rouppe van der Voort, L.: 2012, *Solar Phys.* **280**, 457.
Beckers, J.M.: 1962, *Aust. J. Phys.* **15**, 327.
Beckers, J.M.: 1964, A study of the fine structures in the solar chromosphere, Ph.D. thesis, Univ. Utrecht.
Cao, W., Gorceix, N., Coulter, R., Ahn, K., Rimmele, T.R., Goode, P.R.: 2010, *Astron. Nachr.* **331**, 636.
Chae, J., Park, H.-M., Ahn, K., Yang, H., Park, Y.-D., Nah, J., Jang, B.-H., Cho, K.-S., Cao, W., Goode, P.R.: 2013, *Solar Phys.* **288**, 1.
de Groof, A., Bastiaensen, C., Müller, D.A.N., Berghmans, D., Poedts, S.: 2005, *Astron. Astrophys.* **443**, 319.
del Toro Iniesta, J.C., Bellot Rubio, L.R., Collados, M.: 2001, *Astrophys. J. Lett.* **549**, L139.
Dialetis, D., Mein, P., Alissandrakis, C.E.: 1985, *Astron. Astrophys.* **147**, 93.
Field, G.B.: 1965, *Astrophys. J.* **142**, 531.
Foukal, P.: 1978, *Astrophys. J.* **223**, 1046.
Georgakilas, A.A., Christopoulou, E.B., Skodras, A., Koutchmy, S.: 2003, *Astron. Astrophys.* **403**, 1123.
Goldsmith, D.W.: 1971, *Solar Phys.* **19**, 86.
Goode, P.R., Denker, C.J., Didkovsky, L.I., Kuhn, J.R., Wang, H.: 2003, *J. Korean Astron. Soc.* **36**, 125.
Habbal, S.R., Ronan, R., Withbroe, G.L.: 1985, *Solar Phys.* **98**, 323.
Hasan, S.S., Schuessler, M.: 1985, *Astron. Astrophys.* **151**, 69.
Heinzel, P., Anzer, U.: 2006, *Astrophys. J. Lett.* **643**, L65.
Heinzel, P., Gouttebroze, P., Vial, J.-C.: 1994, *Astron. Astrophys.* **292**, 656.
Ichimoto, K., Shine, R.A., Lites, B., Kubo, M., Shimizu, T., Suematsu, Y., *et al.*: 2007, *Publ. Astron. Soc. Japan* **59**, 593.
Kawaguchi, I.: 1970, *Publ. Astron. Soc. Japan* **22**, 405.
Lemen, J.R., Title, A.M., Akin, D.J., Boerner, P.F., Chou, C., Drake, J.F., *et al.*: 2012, *Solar Phys.* **275**, 17.
Leroy, J.-L.: 1972, *Solar Phys.* **25**, 413.
Louis, R.E., Bellot Rubio, L.R., Mathew, S.K., Venkatakrishnan, P.: 2011, *Astrophys. J.* **727**, 49.
Mein, P., Briand, C., Heinzel, P., Mein, N.: 2000, *Astron. Astrophys.* **355**, 1146.
Müller, D.A.N., Hansteen, V.H., Peter, H.: 2003, *Astron. Astrophys.* **411**, 605.
Müller, D.A.N., Peter, H., Hansteen, V.H.: 2004, *Astron. Astrophys.* **424**, 289.
Narain, U., Ulmschneider, P.: 1990, *Space Sci. Rev.* **54**, 377.
Nicolas, K.R., Kjeldseth-Moe, O.: 1981, In: Cram, L.E., Thomas, J.H. (eds.) *The Physics of Sunspots*, National Solar Observatory, Sunspot, 336.
Parker, E.N.: 1953, *Astrophys. J.* **117**, 431.

Schrijver, C.J.: 2001, *Solar Phys.* **198**, 325.
Tziotziou, K., Heinzel, P., Mein, P., Mein, N.: 2001, *Astron. Astrophys.* **366**, 686.
Vardavas, I.M., Cram, L.E.: 1974, *Solar Phys.* **38**, 367.
Vissers, G., Rouppe van der Voort, L.: 2012, *Astrophys. J.* **750**, 22.
Wallace, L., Hinkle, K., Livingston, W.C.: 1993, *An Atlas of the Photospheric Spectrum from 8900 to 13 600 cm^{-1} (7350 to 11 230 Å)*, National Solar Observatory, Tucson.
Wallace, L., Hinkle, K., Livingston, W.: 1998, *An Atlas of the Spectrum of the Solar Photosphere from 13 500 to 28 000 cm^{-1} (3570 to 7405 Å)*. National Optical Astronomy Observatories, Tucson.
Wang, H., Chae, J., Gurman, J.B., Kucera, T.A.: 1998, *Solar Phys.* **183**, 91.

GPSR Compliance

The European Union's (EU) General Product Safety Regulation (GPSR) is a set of rules that requires consumer products to be safe and our obligations to ensure this.

If you have any concerns about our products, you can contact us on ProductSafety@springernature.com

In case Publisher is established outside the EU, the EU authorized representative is:

Springer Nature Customer Service Center GmbH
Europaplatz 3
69115 Heidelberg, Germany

Batch number: 10584688

Printed by Printforce, the Netherlands